STUDENT SOLUTIONS MANUAL

David S. Rubin
University of North Carolina, Chapel Hill

SEVENTH EDITION

STATISTICS for MANAGEMENT

Richard I. Levin ▪ David S. Rubin

PRENTICE HALL, Upper Saddle River, NJ 07458

Acquisitions Editor: *Thomas Tucker*
Assistant Editor: *Audrey Regan*
Production Editor: *Joseph F. Tomasso*
Manufacturing Buyer: *Paul Smolenski*

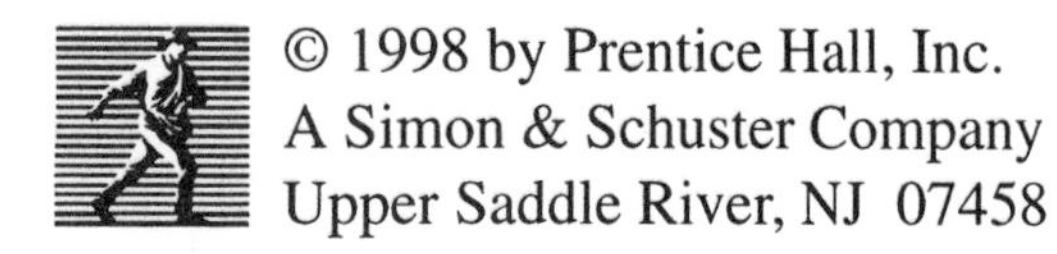

A Simon & Schuster Company
Upper Saddle River, NJ 07458

Printed in the United States of America

10 9 8 7 6 5 4 3 2 1

ISBN 0-13-619636-5

Prentice-Hall International (UK) Limited, *London*
Prentice-Hall of Australia Pty. Limited, *Sydney*
Prentice-Hall Canada Inc., *Toronto*
Prentice-Hall Hispanoamericana, S.A., *Mexico*
Prentice-Hall of India Private Limited, *New Delhi*
Prentice-Hall of Japan, Inc., *Tokyo*
Simon & Schuster Asia Pte. Ltd., *Singapore*
Editora Prentice-Hall do Brasil, Ltda., *Rio de Janiero*

CONTENTS

Preface

This manual provides complete, worked-out solutions to the even- numbered exercises in ***Statistics for Management***, 7th ed., by Richard I. Levin and David S. Rubin. I believe that the best way for you to use these solutions is for checking your own solutions. Work the problems yourself and check your final answer against the answers provided here. If they agree, great! If not, start at the top, compare your procedure with the one here, and see where and how they differ. Use that comparison to understand where you've strayed. (Or where I've strayed—more about that later.) Try to understand what's gone awry, so that the same difficulty won't arise in similar problems in the future.

It's tempting, when worked-out solutions are available, to look at them without first doing your own solutions. The routine goes something like: "Yup, uh-huh, I see that, of course." Unfortunately, that's like reading a recipe in a cook-book and concluding that you can make it well enough to serve to someone important in your life. Until you cook it yourself, you really can't be sure that you'll be able to do it. And if your culinary creation flops, you can go back to the cook-book to try to figure out what went wrong. Chances are that it will come out better the next time.

From personal experience, I realize that one of the greatest frustrations faced by students (and instructors!) of elementary statistics courses is running into mistakes in the brief solutions at the back of the text and in the complete solutions worked out in the Solutions Manual. To minimize that frustration, every effort has been made to make these solutions as error-free as possible. However, it would be both immodest and foolish to assert that both my dedicated typist and I have made no errors in assembling this volume. Fortunately, the wonders of modern technology have enabled us to place the entire manuscript on a small number of disks, and it is easy to make corrections. I hope that the entire *Statistics for Management* package has helped make statistics relevant and comprehensible, and I'd like to ask your help in making it even better for future users. If, as I am afraid inevitably must happen, you encounter any errors in these solutions, please drop me a note at the address below. In turn, I will correct these solutions and send you new pages to replace the incorrect ones. These corrected pages will also be sent to Prentice Hall for circulation to all adopters of *Statistics for Management.* And I'll be pleased to acknowledge your assistance in the next edition of this manual. Many thanks for your understanding, forebearance, and assistance.

David S. Rubin
CB#3490, McColl Building
University of North Carolina
Chapel Hill, NC 27599-3490

CHAPTER 1

INTRODUCTION

There are NO problems
in Chapter 1

Please move on
to Chapter 2

CHAPTER 2

ARRANGING DATA TO CONVEY MEANING: TABLES AND GRAPHS

2-2 Since the Department of Commerce keeps statistics on all the cars sold in the U.S., this conclusion is drawn from a population.

2-4 On the basis of German history since the end of World War II, and given the bias produced by his own strong belief in the validity of Communism, Ulbricht was unable to foresee the possibility of the changes that resulted from Gorbachev's hands-off policy toward the eastern European satellite nations.

2-6 The point of this section is to show that the raw data are not in a useful form. We cannot draw any conclusions from these data in their current form. We would first need to do a certain amount of rearranging, such as listing the grades from highest to lowest or determining the most frequent grade pair, before we could begin to make meaningful observations about the problem at hand.

2-8 Here is a clear-cut case of data which has already undergone statistical analysis. In this case, the raw data would be a list of sample units indicating whether or not they were defective. The quality control section has already performed an analysis on these data to calculate the averages contained in the report. Anytime you are given "averages" or "rates" or "highest/lowest" figures, you are dealing with data which have already been statistically reduced.

2-10 In addition to the 7 stores with under 475 service actions which are not breaking even, another 6 stores fall on the "store watch list."

2-12

5 Intervals

Class	Frequency	Relative Frequency
15 - 25	3	0.0667
26 - 36	4	0.0889
37 - 47	12	0.2667
48 - 58	18	0.4000
59 - 69	8	0.1778
	45	1.0000

11 Intervals

Class	Frequency	Relative Frequency
15 - 19	3	0.0667
20 - 24	0	0.0000
25 - 29	2	0.0444
30 - 34	2	0.0444
35 - 39	5	0.1111
40 - 44	3	0.0667
45 - 49	11	0.2444
50 - 54	3	0.0667
55 - 59	8	0.1778
60 - 64	3	0.1111
65 - 69	3	0.0667
	45	1.0000

a) No: we can see from either distribution that more than 10% of the motorists drive at 55 mph or more. (5 intervals $\Rightarrow$ over 17.78%; 11 intervals $\Rightarrow$ 35.56%)

b) Either can be used; the 11-interval distribution gives a more precise answer.

c) The 5-interval distribution shows that 66.67% of the motorists drive between 37 and 58 mph inclusive.

2-14 Sorting the data in ascending order by number of channels purchased gives us:

Channels purchased	17	18	22	25	28	29	39	42	43	76	84	96	104
Hours watched	19	16	13	14	13	7	9	12	16	8	4	6	6

This shows that hours watching television decreases as number of channels purchased increases. Sorting by number of hours watched leads to the same conclusion.

2-16 a) First we will construct the raw data from the information in the problem.

Spread	SAT Differential	Spread	SAT Differential
− 1.1	− 120	− 0.2	20
0.1	− 20	0.1	− 10
− 0.5	− 30	0.1	20
− 0.5	− 90	0.3	50
0.2	0	0.4	60
0.1	− 10	0.6	60
0.3	60	0.0	10
1.3	140	− 0.6	− 120
− 0.2	0	− 0.7	− 100
− 0.1	− 10	0.8	150

Now, we will array these data by highest to lowest spreads.

Spread	SAT Differential	Spread	SAT Differential
1.3	140	0.1	− 20
0.8	150	0.0	10
0.6	60	− 0.1	− 10
0.4	60	− 0.2	0
0.3	60	− 0.2	0
0.3	50	− 0.5	− 30
0.2	20	− 0.5	− 90
0.1	20	− 0.6	− 120
0.1	− 10	− 0.7	− 100
0.1	− 10	− 1.1	− 120

b) From the array we can see that the most common spread is 0.1, which occurs four times.

c) For a spread of 0.1, the most common SAT differential is − 10, which occurs twice out of the four data points.

d) The SAT differential appears to be a good indicator of spread: students with high SAT differentials do better in college than in high school, and students with large negative differentials do worse.

2-18

Classes (lbs/sq in)	Relative Frequencies
2490.0 - 2493.9	.150
2494.0 - 2497.9	.175
2498.0 - 2501.9	.325
2502.0 - 2505.9	.225
2506.0 - 2509.9	.125
	1.000

The greatest number of these samples (32.5%) fell into the class 2498.0 - 2501.9 lbs/sq in. This would have been somewhat difficult to see from the data in the table.

2-20 a)

"Before" Classes	Frequency	Relative Frequencies
1 to 2	5	.25
3 to 4	6	.30
5 to 6	7	.35
7 to 8	2	.10
9 to 10	0	.00
	20	1.00

b)

"After" Classes	Frequency	Relative Frequencies
1 to 2	2	.10
3 to 4	4	.20
5 to 6	6	.30
7 to 8	6	.30
9 to 10	2	.10
	20	1.00

c) In order to be able to make comparisons of the frequency distributions

d)

"Change" Classes	Frequency	Relative Frequencies
− 5 to − 4	1	.05
− 3 to − 2	0	.00
− 1 to 0	5	.25
1 to 2	8	.40
3 to 4	5	.25
5 to 6	1	.05
	20	1.00

e) Sales appear to have increased, but the apparent increase could be due to other factors we don't know about, so we can't say for sure that the new slogan has helped.

2-22

Class	< 25	25-34	35-44	45-54	≥55
Frequency	6	9	7	3	5
Relative Frequency	.200	.300	.233	.100	.167

a) Now we see that most purchasers are under 45.

b) We can be more precise: about 75% of the purchasers are under 45.

2-24 a) No. Since most of the runners will take between 5 and 6 minutes per mile, almost all of the observations will fall into the two lower classes. As a result, the coach won't get much information from this distribution.

b) Five classes with midpoints at 25, 27, 29, 31, and 33 would be much more helpful.

2-26 To construct a closed classification, we must insure that the list is all-inclusive. A distribution containing the following categories would meet this requirement: Single, Married, Divorced, Separated, Widowed.

To construct an open-ended classification, we can use any of the categories listed above plus the category "other". Since the problem asks for only three categories, we can choose any two of the above. The logical choice would be to use the following: Single, Married, Other

Notice that as you open the distribution you lose completeness, but you gain simplicity.

2-28 The classes run from 85 to 114 and from 115 to 144, with additional open-ended classes at the bottom (84 and under) and at the top (145 and over). Since the group wants to highlight the noisy flights, this distribution is inadequate, because the 115 to 144 class includes noise levels on both sides of the 140 decibel limit.

distribution is inadequate, because the 115 to 144 class includes noise levels on both sides of the 140 decibel limit.

2-30 Looking at the data, we see that 22 days is the minimum and 51 days is the maximum. Using equation 2-1, the intervals will be

$$\frac{51 - 22}{10} = 2.9\text{, or approximately 3 days wide for part (a)}$$

and

$$\frac{51 - 22}{5} = 5.8\text{, or approximately 6 days wide for part (b).}$$

a)

Waiting Time (days)	Frequency
22 - 24	3
25 - 27	3
28 - 30	6
31 - 33	12
34 - 36	8
37 - 39	6
40 - 42	5
43 - 45	4
46 - 48	2
49 - 51	1
	50

b)

Waiting Time (days)	Frequency
22 - 27	6
28 - 33	18
34 - 39	14
40 - 45	9
46 - 51	3
	50

With ten divisions the 31-33 interval has the highest number of observations with 12; however, when only five divisions are used, then the 28-33 interval has the most observations with 18.

c) Yes, since he wants to know the relative proportions at each level.

2-32 a) Discrete and closed

b) Discrete and closed

c) Flavor is qualitative, amount is quantitative

d) He should collect data on how often the stores ran out of the various flavors and how much of each flavor was left over.

2-34

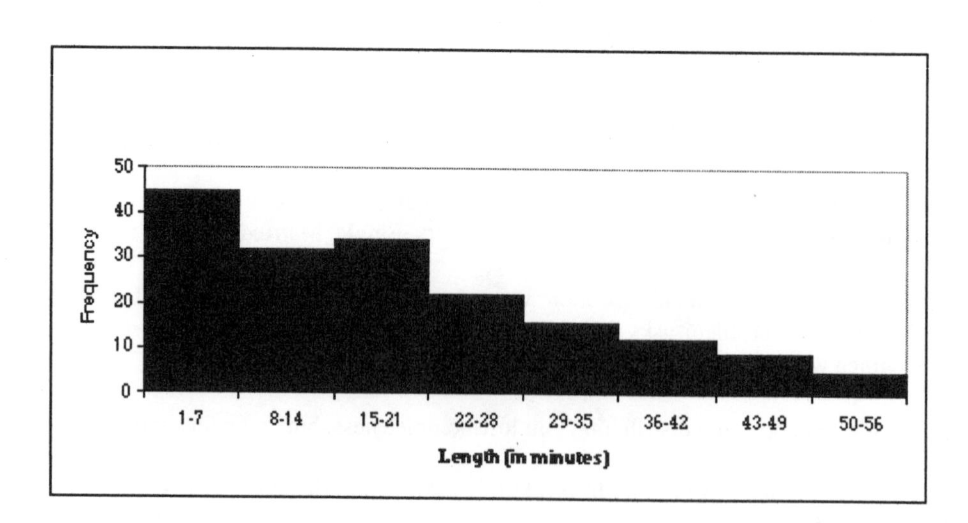

a) With one minor exception (the 15-21 minute interval), the frequencies decrease as the length of the calls increases.

b) Let the people with the longest calls go first. Whenever a phone becomes available, let a person in the highest class with calls still remaining make the next call.

c) Yes. If we allowed all the 1-7 minute calls to be made first and then worked upward, the 50-56 minute calls would have to wait until the shorter calls were made. Using the order suggested in (b) can save time by allowing shorter calls to be made while longer calls are in progress.

2-36

Class	Frequency	Cumulative Relative Frequency
2000 - 3999	3	.15
4000 - 5999	7	.50
6000 - 7999	7	.85
8000 - 9999	3	1.00

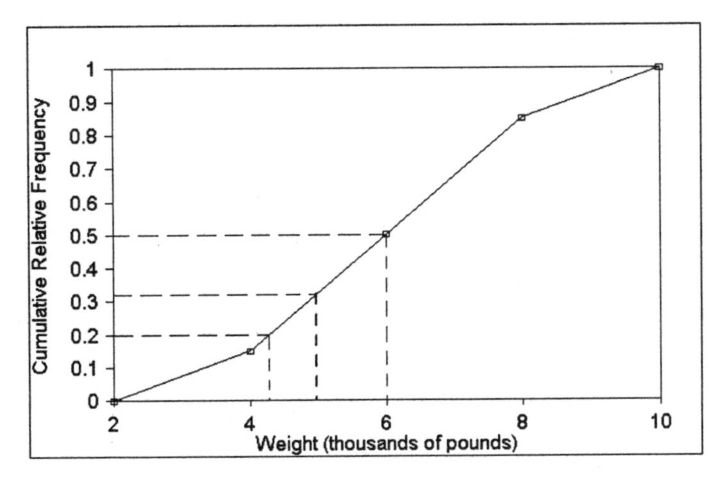

a) About 65% of all trips break even by catching at least 5000 pounds.

b) The middle value is approximately 6000 pounds.

c) 80% of the catches are greater than approximately 4300 pounds.

2-38 a)

River Flow (>)	1000	1050	1100	1150	1200	1250	1300	1350	1400
Cum. Frequency	246	239	218	186	137	79	38	11	0

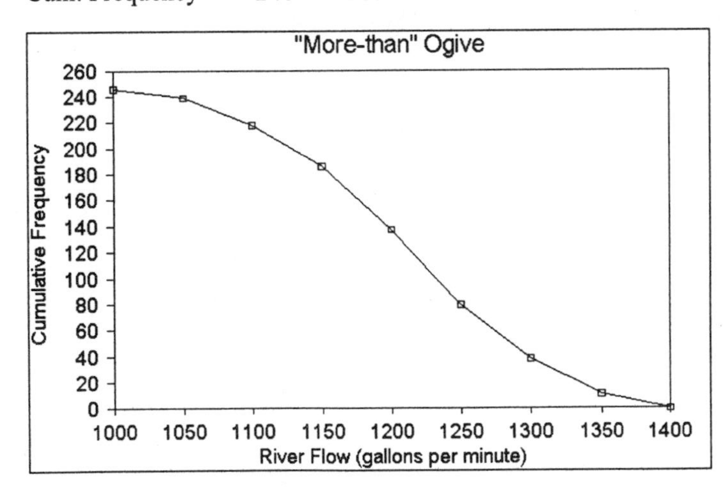

b)

River Flow (<)	1001	1051	1101	1151	1201	1251	1301	1351	1401
Cum. Frequency	0	7	28	60	109	167	208	235	246

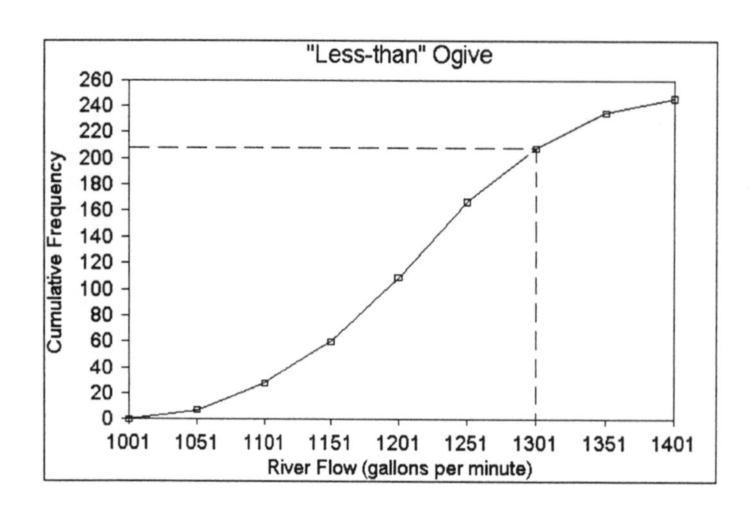

c) About 85%.

2-40 a)

19.0	20.7	21.8	23.1	24.1
19.5	20.8	21.9	23.3	24.1
19.5	20.9	22.0	23.5	24.2
19.7	20.9	22.2	23.6	24.2
19.8	20.9	22.5	23.7	24.2
19.9	21.1	22.7	23.8	24.3
20.1	21.2	22.8	23.8	25.0
20.3	21.3	22.8	23.8	25.0
20.7	21.5	22.8	23.9	25.1
20.7	21.6	22.9	23.9	25.3

b)

Minutes to Set Type	Frequency	Minutes to Set Type ($\leq$)	Frequency
19.0 - 19.7	4	19.0	0
19.8 - 20.5	4	19.8	4
20.6 - 21.3	10	20.6	8
21.4 - 22.1	5	21.4	18
22.2 - 22.9	7	22.2	23
23.0 - 23.7	5	23.0	30
23.8 - 24.5	11	23.8	35
24.6 - 25.3	4	24.6	46
	50	25.4	50

c)

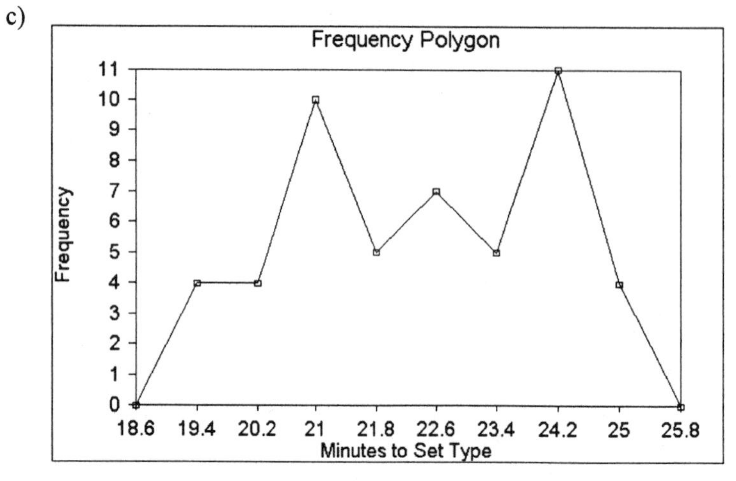

d)

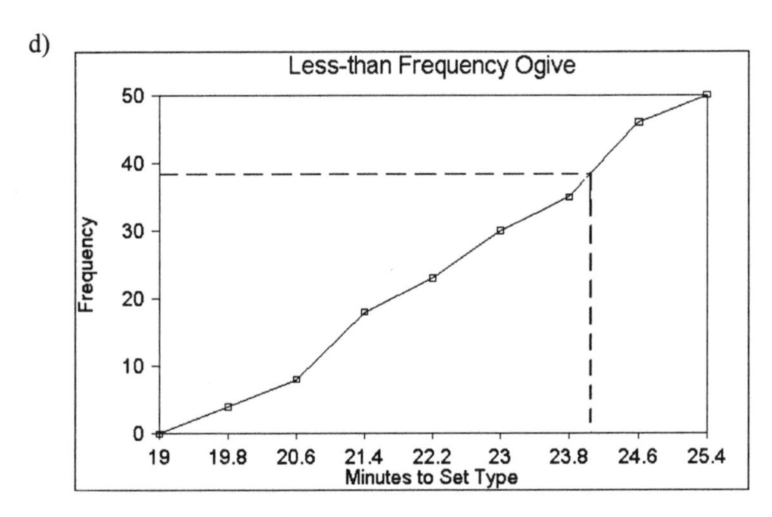

e) About 78% of the time.

2-42

Class ($)	Cumulative Relative Frequency	Class ($)	Cumulative Relative Frequency
< 5,001	.038	20,001 - 30,000	.731
5,001 - 10,000	.108	30,001 - 40,000	.877
10,001 - 15,000	.192	40,001 - 50,000	.946
15,001 - 16,000	.446	> 50,000	1.000

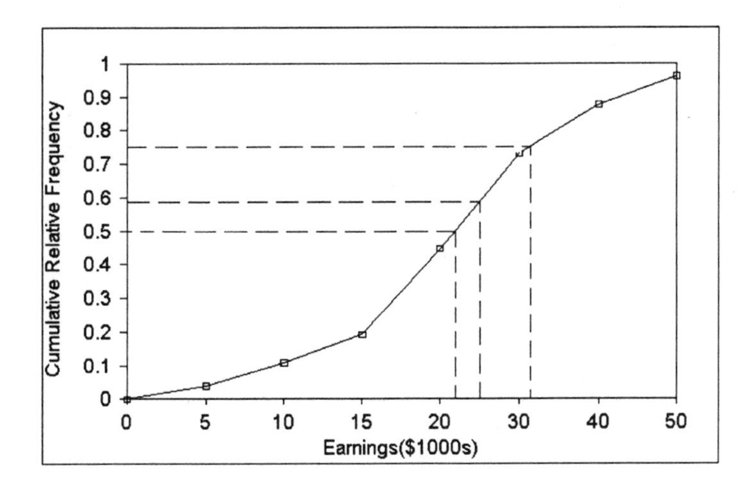

a) About 42%.

b) About $22,000.

c) About $31,000.

2-44 By grouping those of the same educational level together, we can more clearly see group differences associated with educational level.

Educational Level	Salary range
Did not finish high school	$14,400 - 17,600
High school graduates	17,000 - 30,400
One or more years of college	14,400 - 22,400
College graduates	19,600 - 34,400
Master's degree	23,200 - 36,200
PhD degrees	29,000 - 64,000
Doctors & lawyers	52,000 - 100,000

2-46 a)

1.9	1.8	1.7	1.6	1.5	1.5	1.5	1.5	1.2	0.9
0.9	0.9	0.9	0.8	0.7	0.7	0.5	0.4	0.4	0.3

b)

Class (inches)	Relative Frequency	Cumulative Relative Frequency
0.000 - 0.249	.00	.00
0.250 - 0.499	.15	.15
0.500 - 0.749	.15	.30
0.750 - 0.999	.25	.55
1.000 - 1.249	.05	.60
1.250 - 1.499	.00	.60
1.500 - 1.749	.30	.90
1.750 - 1.999	.10	1.00

c) The data are distinctly bimodal, with modal classes 0.750 - 0.999 and 1.500 - 1.749.

d)

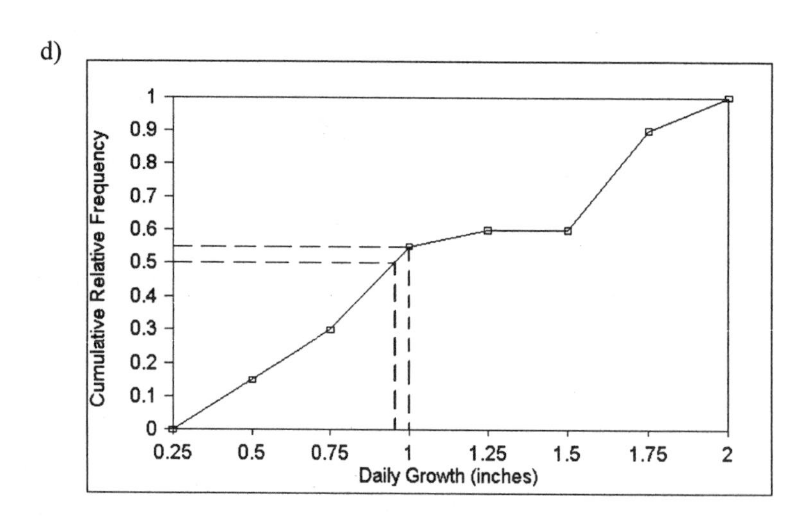

About 45% grew more than 1.0 inches per week.

e) About .95 inches.

2-48

Class (units)	Cumulative Relative Frequency
9700 - 9899	.200
9900 - 10099	.733
10100 - 10299	.867
10300 - 10499	.867
10500 - 10699	1.000

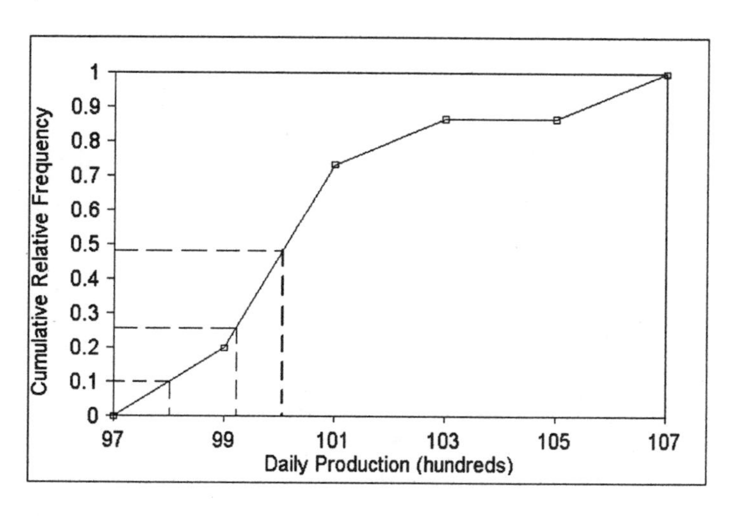

a) About 50% (7 or 8 items) exceeded the breakeven point.

b) About 9,900 units.

c) About 9,800 units.

2-50 It tells you what fraction of the observations fit into each class. This makes it easier to compare samples or populations of different sizes.

2-52 $\frac{2000}{2000+8000} = .2$, so there should be $.2 \times 250 = 50$ women and 200 men.

2-54

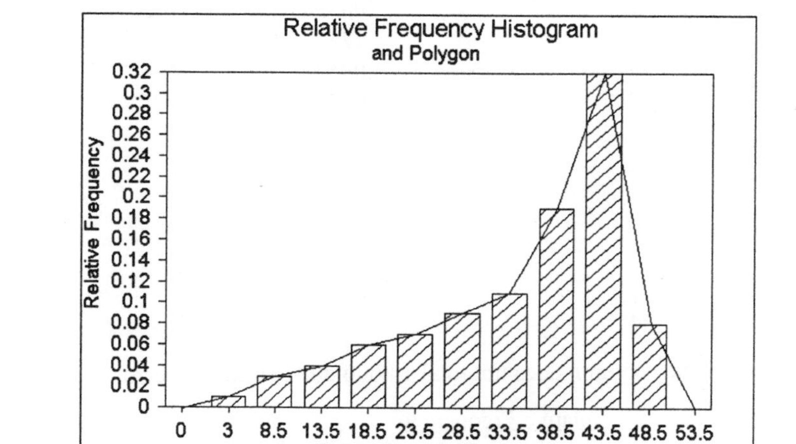

2-56

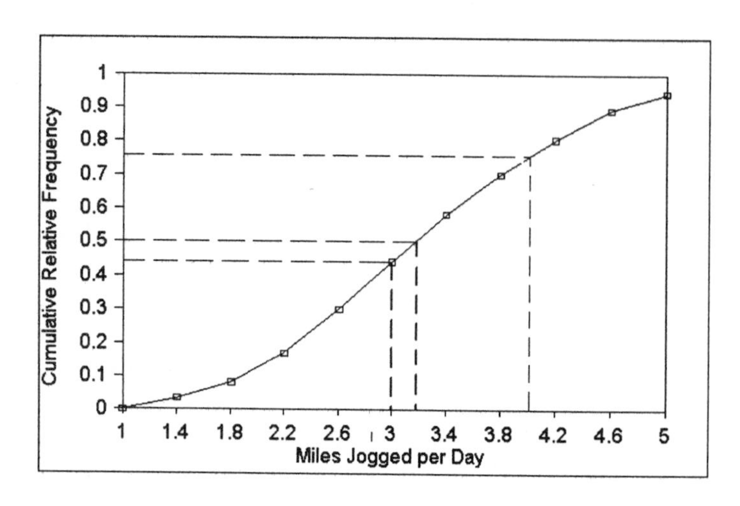

About 75% of the joggers average 4 or fewer miles per day.

2-58 Since the Ivy League schools have more male than female undergraduates, this is not a representative sample.

2-60 a)

Journal Number	Frequency	Relative Frequency
1	1	.0417
2	2	.0833
3	2	.0833
4	0	.0000
5	2	.0833
6	2	.0833
7	3	.1250
8	1	.0417
9	2	.0833
10	1	.0417
11	2	.0833
12	2	.0833
13	0	.0000
14	2	.0833
15	2	.0833
	24	.9998

NOTE: The relative frequency column does not total 1.0000 due to rounding errors.

b)

Branch	Frequency	Relative Frequency
North	11	.4583
West	8	.3333
South	5	.2083
	24	.9999

c)

# of Publications	Frequency	Relative Frequency
1 - 3	6	.2500
4 - 6	5	.2083
7 - 9	4	.1667
10 - 12	4	.1667
13 - 15	2	.0833
16 - 18	2	.0833
19 - 21	1	.0417
	24	1.0000

d) Although the faculty use of journals is widespread, it is clear that the North branch accounts for most of the publications. Well over half of the faculty (about 62%) publish 9 or fewer articles.

2-62

1) Quantitative, discrete, open-ended: in all likelihood there would be open-ended classes like under 20 and over 50.

2) Quantitative, discrete, open-ended: people will report incomes to the nearest thousand dollars, and for very high and low incomes, the company won't care about precise values.

3) Qualititative, discrete, closed: the only possibilities are single, married, divorced, and widowed.

4) and 5) Both of these distributions would be qualitative and discrete. However, in contrast to 3), the answers could be so varied that it is unlikely that the company would choose to list all possibilities. Instead, the most frequent responses would be used and "other" included to cover all other possibilities. Therefore, these distributions would most likely be open-ended.

2-64

Group I

None	Mild	Moderate	Severe	or	None	- 3
None	Mild	Moderate	Severe		Mild	- 7
None	Mild	Moderate	Severe		Moderate	- 5
	Mild	Moderate			Severe	- 3
	Mild	Moderate				
	Mild					
	Mild					

Group II

None	Mild	Moderate	Severe	or	None	- 1
	Mild	Moderate	Severe		Mild	- 4
	Mild	Moderate	Severe		Moderate	- 8
	Mild	Moderate	Severe		Severe	- 5
		Moderate	Severe			
		Moderate				
		Moderate				
		Moderate				

Displayed this way, it is much easier to compare the two groups.

2-66

The problem asks for a distribution by specialty, not combinations. Therefore, the only categories would be accounting, marketing, statistics, finance, and no publications. The faculty who have more than one specialty will be double counted in this particular distribution.

Specialty	Frequency	Relative Frequency
Accounting	17	.140
Marketing	41	.339
Statistics	40	.331
Finance	22	.182
No Publications	1	.008
	121	1.000

2-68

a, b)

Percentage won	Frequency	Relative Frequency	Cumulative Relative Frequency
0.000 - 0.199	2	0.0714	0.0714
0.200 - 0.399	8	0.2857	0.3571
0.400 - 0.599	7	0.2500	0.6071
0.600 - 0.799	10	0.3571	0.9643
0.800 - 0.999	1	0.0357	1.0000

c)

d) See part (a) for the cumulative relative frequencies.

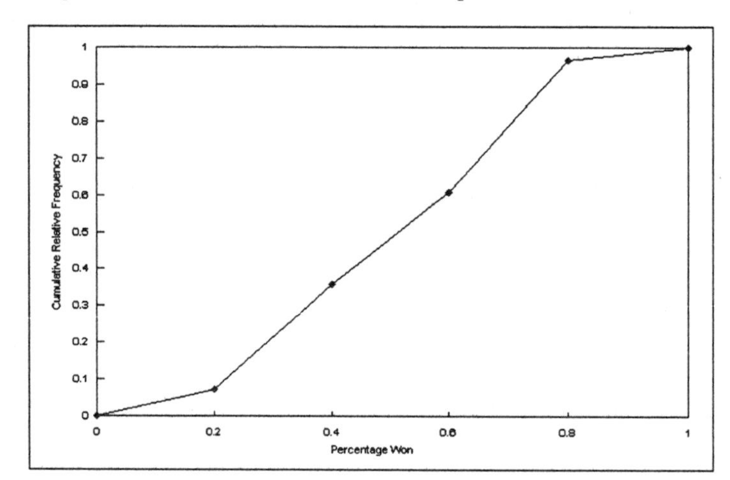

e) The one team in the 0.800 - 0.999 class gets a playoff berth, as do 5 of the 10 teams in the 0.600 - 0.799 class.

2-70 a) 139 129 128 126 121 119 119 116 115 114 113 113 112 111 110
110 108 107 105 102 101 100 99 99 97 93 93 91 87 84
80 75 72 66 60

b) 16 exceeded the limit of 108 minutes, 18 were under, 1 was exactly at the limit.

c)

Class (min.)	60-69	70-79	80-89	90-99	100-109	110-119	120-129	130-139
Frequency	2	2	3	6	6	11	4	1
Relative Freq.	.057	.057	.086	.171	.171	.314	.114	.029

d) If 108 minutes is typical, then about half should be above 108 and half below. The data support this. Since we don't know how much downtime per shift is viewed as excessive, we cannot tell if Cline should be concerned or not.

2-72 a) 151.1 147.8 145.7 142.3 142.0 142.0 141.2 141.1 140.9 138.7
138.2 137.4 134.9 133.3 133.0 130.8 129.8 128.9 126.3 125.7
125.7 125.2 125.0 119.9 118.6

b) 23/25 (or 92%) withstood 120,000 pounds of force; only 1/25 (or 4%) withstood 150,000 pounds of force.

c) 16/25 (or 64%) would have failed. They should stop ordering from this supplier until it can improve the strength of its bolts.

2-74 a) 9, 8, 7, 7, 6, 6, 5, 5, 5, 5, 4, 3, 3, 3, 3, 2, 2, 2, 1, 1, 1, 1, 1, 1, 1, 1, 1, 1

b) 1.

Class (sales)	1	2	3	4	5	6	7	8	9
Frequency	10	3	4	1	4	2	2	1	1
Relative Frequency	.357	.107	.143	.036	.143	.071	.071	.036	.036

2.

Class (sales)	1-3	4-6	7-9
Frequency	17	7	4
Relative Frequency	.607	.250	.143

Both distributions are skewed: many countries have relatively few sales, and then the distribution tails off to the right.

2-76 a) 4600 - 5199, 5200 - 5799, 5800 - 6399, 6400 - 6999, 7000 - 7599, and 7600 - 8199.

b) 0.00 - 1.39, 1.40 - 2.79, 2.80 - 4.19, 4.20 - 5.59, 5.60 - 6.99, and 7.00 - 8.39.

2-78 a) 8/61 = 13.1% b) 29/61 = 47.5%

2-80 A histogram will further highlight the pattern of low numbers of customers from 11 p.m. to 6 a.m., then increasing fairly steadily until 11 a.m. or noon, remaining reasonably steady until 5 p.m., and then decreasing fairly steadily until 11 p.m. One limitation: national data may not apply to Utah. For example, if there were many factories close by, the number of customers at Fresh Foods might reflect changes in work shifts at those factories.

CHAPTER 3

MEASURES OF CENTRAL TENDENCY AND DISPERSION IN FREQUENCY DISTRIBUTIONS

3-2

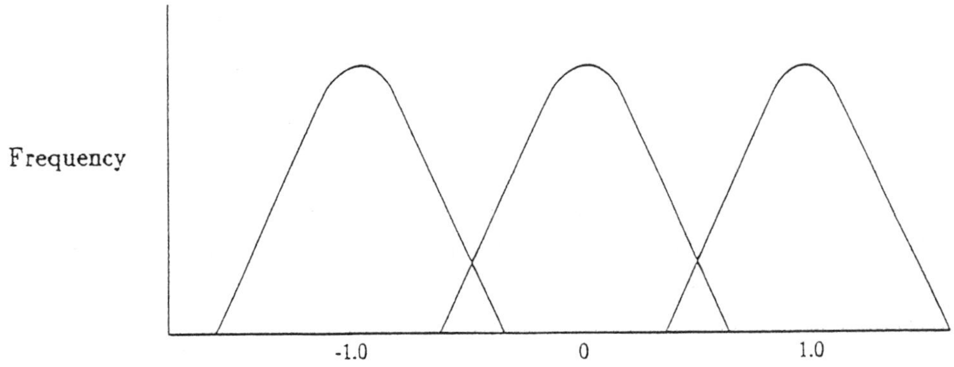

Frequency

-1.0 0 1.0

3-4 a) B b) A c) A d) B e) B f) A g) neither

3-6 $\bar{x} = (8 + 5 + 9 + 10 + 9 + 12 + 7 + 12 + 13 + 7 + 8)/11 = 100/11 = 9.091$
Since this is over 9, they do not qualify.

3-8 a)

Age	40 - 49	50 - 59	60 - 69	70 - 79	80 - 89
Frequency	4	4	3	2	7

b) $\bar{x} = \dfrac{\sum(f \times x)}{n} = \dfrac{4(45) + 4(55) + 3(65) + 2(75) + 7(85)}{20} = \dfrac{1340}{20} = 67$

c) $\bar{x} = \dfrac{\sum x}{n} = \dfrac{1335}{20} = 66.75$

d) As expected, they are close, but not exactly the same.

3-10 For the first six months, the average is

$(234 + 216 + 195 + 400 + 315 + 274)/6 = 1634/6 = 272.33$ animals.

For the entire year, the average is

$(1634 + 302 + 291 + 275 + 300 + 375 + 450)/12 = 3627/12 = 302.25$ animals.

Since the criterion is not met for the first six months, the owner will not build the new store.

3-12 $\bar{x} = \dfrac{\sum x}{n} = \dfrac{465.9}{20} = 23.295.$
Since this exceeds 23, the manager should be concerned.

3-14

	Q_1	Q_2	Q_3	Q_4	(b)
Y_1	10	5	25	15	13.75
Y_2	20	10	20	10	15.00
Y_3	30	15	45	50	35.00
(a)	20	10	30	25	21.25 (c)

(All figures in thousands of dollars)

3-16

Student 1: $(.20 \times 85) + (.10 \times 89) + (.10 \times 94) + (.25 \times 87) + (.35 \times 90) = 88.55$
Student 2: $(.20 \times 78) + (.10 \times 84) + (.10 \times 88) + (.25 \times 91) + (.35 \times 92) = 87.75$
Student 3: $(.20 \times 94) + (.10 \times 88) + (.10 \times 93) + (.25 \times 86) + (.35 \times 89) = 89.55$
Student 4: $(.20 \times 82) + (.10 \times 79) + (.10 \times 88) + (.25 \times 84) + (.35 \times 93) = 86.65$
Student 5: $(.20 \times 95) + (.10 \times 90) + (.10 \times 92) + (.25 \times 82) + (.35 \times 88) = 88.50$

3-18

$$\bar{x}_w = \frac{\sum(w \times x)}{\sum w} = \frac{0(897) + 1(1082) + 2(1325) + 3(814) + 4(307) + 5(253) + 6(198)}{897 + 1082 + 1325 + 814 + 307 + 253 + 198}$$
$$= \frac{9855}{4876} = 2.021 \text{ times}$$

3-20

$$\bar{x}_w = \frac{\sum(w \times x)}{\sum w}$$
$$= \frac{16400(.05) + 24100(.08) + 77600(.13) + 1900(.17) + 1300(.35) + 750(.40) + 800(.45)}{16400 + 24100 + 77600 + 1900 + 1300 + 750 + 800}$$
$$= \frac{14274}{122850} = \$0.1162 \text{ per ounce}$$

3-22

$GM = \sqrt[5]{1.05(1.105)(1.09)(1.06)(1.075)} = \sqrt[5]{1.441094314} = 1.0758172$

So the average increase is about 7.58% per year.

3-24

$GM = \sqrt[4]{17630/12500} = \sqrt[4]{1.4104} = 1.0897718$

So the average increase is 8.98% per year. In 1996, the estimated production is $17630(1.0898)^3 = 22819$ units.

3-26

$GM = \sqrt{1.15/1.00} = 1.0723805.$

So the price has been increasing by about 7.24% per week.

3-28

$GM = \sqrt[4]{66/55} = \sqrt[4]{1.2} = 1.0466351$

So the average increase is about 4.66% per year. In three more years, the estimated cost is $66(1.0466)^3 =$ \$75.66.

3-30

We first arrange the mileages in ascending order:

210	447	450	469	488	559	560	589	657	689
756	775	788	789	810	876	890	943	987	1450

a) median $= \frac{689 + 756}{2} = 722.5$ miles, the average of items 10 and 11

b) $\bar{x} = \frac{\sum x}{n} = \frac{14182}{20} = 709.1$ miles

c) In this instance, both are equally good since they are so close to each other.

3-32

Class	Frequency	Cumulative Frequency	Class	Frequency	Cumulative Frequency
10 - 19.5	8	8	60 - 69.5	52	181
20 - 29.5	15	23	70 - 79.5	84	265
30 - 39.5	23	46	80 - 89.5	97	362
40 - 49.5	37	83	90 - 99.5	16	378
50 - 59.5	46	129	≥100	5	383

a) The median is the (383 + 1)/2 = 192nd item.

b) The median class is 70 – 79.5.

c) The step width in the median class = 10/84 = 0.1190

d) median = 70 + 10(.1190) = 71.190

e) $\tilde{m} = \left(\frac{(n+1)/2-(F+1)}{f_m}\right)w + L_m = \left(\frac{192-82}{84}\right)10 + 70 = 71.1905$

The answers in (d) and (e) differ slightly because of rounding

3-34 We first arrange the times in ascending order:

17	19	21	22	22	28	29	29	29	30
32	33	33	34	34	39	41	43	44	52

The median time is (30 + 32)/2 = 31 minutes, the average of the 10th and 11th items. This is close enough to the target of 30 minutes to conclude that excessive speeds have not been a problem.

3-36 The median is the (4723 + 1)/2 = 2362nd element, which occurs as the 907th element in the 750 - 999.99 class. The distance between elements in this class is 250/1776 = .14077, so the median is 750 + 906(.14077) = \$877.54. Using equation 3-8, we have

$$\tilde{m} = \left(\frac{(n+1)/2-(F+1)}{f_m}\right) w + L_m = \left(\frac{(4723+1)/2-(1456)}{1776}\right) 250 + 750$$

$$= \$877.54$$

3-38

Books checked out (x)	0	1	2	3	4	5	6	7	
Frequency (f)	3	3	7	3	2	1	0	1	
$f \times x$	0	3	14	9	8	5	0	7	$\sum f \times x = 46$

a) The mode is 2 books.
b) The mean is 46/20 = 4.3 books.
c) Since the distribution is skewed, the mode is the better measure of central tendency.

3-40 a) Brunette b) A c) Wednesday and Saturday

3-42 $\text{Mo} = L_{\text{Mo}} + \frac{d_1}{d_1+d_2} w = 750 + \left(\frac{710}{710+284}\right) 250 = \$928.57$

3-44 a) The modal class is \$1000 - \$1499

b) $L_{\text{Mo}} = 1000$ $d_1 = 96$ $d_2 = 104$ $w = 500$

$$\text{Mo} = L_{\text{Mo}} + \frac{d_1}{d_1+d_2} w = 1000 + \left(\frac{96}{96+104}\right) 500 = \$1240$$

c) .90(1240) = \$1116. All 535 applicants who earn below \$1000 qualify. If we assume the applicants in the 1000 - 1499 class are evenly distributed within that class, then about (116/500)(400) = 93 of them qualify as well. Thus about 628 applicants are qualified.

3-46 C

3-48 A, because it has less variability

3-50 There are many ways that the concept may be involved. Certainly, the FTC would be examining the price variability for the industry and comparing the result to that of the suspect companies. The agency might examine price distributions for similar products, for the same products in a city, or for the same products in different cities. If the variability was significantly different in any of these cases, this result might constitute evidence of a conspiracy to set prices at the same levels.

3-52 First we arrange the data in increasing order:

33	45	52	54	55	- 1st quartile
61	66	68	69	72	- 2nd quartile
74	75	76	77	84	- 3rd quartile
91	91	93	97	99	- 4th quartile

Interquartile range $= Q_3 - Q_1 = 84 - 55 = 29$

3-54 Placing the 30 temperatures in ascending order

69	72	82	84	84	86	87	87	88	88	88	88	89	89	89
92	92	94	94	94	94	95	96	97	98	99	99	102	102	105

we see that the 70th percentile (the 21st observation) is 94 degrees.

3-56 First we arrange the data in increasing order:

.10	.12	.23	.32	.45 - 1st quartile
.48	.50	.51	.53	.58 - 2nd quartile
.59	.66	.67	.69	.77 - 3rd quartile
.89	.95	.99	1.10	1.20 - 4th quartile

Range = 1.20 − .10 = 1.10 minutes
Interquartile range $= Q_3 - Q_1 = .77 - .45 = .32$ minutes

3-58 Range = highest − lowest = 502.6 − 6.3 = 496.3 megabytes
Interquartile range $= Q_3 - Q_1 = 405.6 - 29.5 = 376.5$ megabytes.

3-60 First we arrange the data in increasing order:

43 57 104 162 201 220 253 302 380 467 500 633

a) 2nd decile = 57 (2nd smallest data value)
8th decile = 467 (10th smallest data value)
Interfractile range = 467 − 57 = 410

b) Median = (220 + 253)/2 = 236.5
$Q_1 = 104$ (3rd smallest value)
$Q_3 = 380$ (9th smallest value)

c) Interquartile range $= Q_3 - Q_1 = 380 - 104 = 276$

3-62

x	17	21	18	27	17	21	20	22	18	23
x^2	289	441	324	729	289	441	400	484	324	529

$\sum x = 204 \quad \sum x^2 = 4250$

$\bar{x} = \sum x/n = 204/10 = 20.4$

$s^2 = (\sum x^2 - n\bar{x}^2)/(n - 1) = (4250 - 10(20.4)^2)/9 = 9.8222$

$s = \sqrt{s^2} = \sqrt{9.8222} = 3.1340$

This exceeds the allowable variability of 3 boats per day; she should be concerned.

3-64

Class	Midpoint (x)	Frequency (f)	$f \times x$	$x - \mu$	$(x - \mu)^2$	$f(x - \mu)^2$
0 - 199	100	10	1000	– 490	240100	2401000
200 - 399	300	13	3900	– 290	84100	1093300
400 - 599	500	17	8500	– 90	8100	137700
600 - 799	700	42	29400	110	12100	508200
800 - 999	900	18	16200	310	96100	1729800
		100	59000			5870000

$$\mu = \frac{\sum f \times x}{N} = \frac{59000}{100} = 590 \text{ checks per day}$$

$$\sigma^2 = \frac{\sum f(x - \mu)^2}{N} = \frac{5870000}{100} = 58700$$

$\sigma = \sqrt{58700} = 242.28$ checks per day. Since this is > 200, Hank should worry.

3-66

Class	Midpoint (x)	Frequency (f)	$f \times x$	$x - x$	$(x - \bar{x})^2$	$f(x - \bar{x})^2$
1 - 3	2	18	36	– 5.715	32.6612	587.9021
4 - 6	5	90	450	– 2.715	7.3712	663.4102
7 - 9	8	44	352	0.285	0.0812	3.5739
10 - 12	11	21	231	3.285	10.7912	226.6157
13 - 15	14	9	126	6.285	39.5012	355.5110
16 - 18	17	9	153	9.285	86.2112	775.9010
19 - 21	20	4	80	12.285	150.9212	603.6849
22 - 24	23	5	115	15.285	233.6312	1168.1561
		200	1543			4384.7550

a) $\bar{x} = \frac{\sum f \times x}{n} = \frac{1543}{200} = 7.715$ days

$$s^2 = \frac{\sum f(x - \bar{x})^2}{n - 1} = \frac{4384.755}{199} = 22.0339, \text{ so } s = \sqrt{22.0339} = 4.69 \text{ days}$$

b) The interval 0 to 17 is roughly the mean $\pm$ 2 standard deviations, so about 75% of he data, or .75(200) = 150 observations should fall in the interval. In fact something between 182 and 191 of the observations are in the interval.

c) About 95% or .95(200) = 191 stays can be expected to fall in the interval from 0 to 17.

3-68 First, we will calculate the standard deviation for the distribution:

$$\sigma = \sqrt{\sigma^2} = \sqrt{49{,}729} = 223$$

A production of 11,175 loaves is one standard deviation below the mean (11,398 – 11,175) = 223. Assuming that the distribution is symmetrical, we know that within $\pm$ one standard deviation from μ fall about 68% of all observations. The interval from the mean to one standard deviation below the mean would contain about 34% (68% ÷ 2) of the data. Therefore, 50% − 34% = 16% (or approximately 5 weeks) of the data would be below 11,175 loaves.

A similar argument can be made for two standard deviations above the mean. Again, assuming a symmetrical distribution, 47.5% of the data (95% ÷ 2) would be contained within this range. This leaves 50% − 47.5% = 2.5%, or about one week (.025 × 32 = .8) with production above 11,844 loaves.

3-70 The key to this problem is to express the actual deviations in terms of standard scores. For example, the first product response latency is off

$$\frac{2.495 - 2.500}{.004} = \frac{-.005}{.004} = -1.25 \text{ standard scores.}$$

For the other products,

$$\text{II: } \frac{2.790 - 2.800}{.006} = \frac{-.010}{.006} = -1.67 \qquad \text{III: } \frac{3.900 - 3.700}{.09} = \frac{.20}{.09} = 2.22$$

Disregarding the signs, we can see that product III produced the response latency with the largest deviation.

3-72 $\sigma^2 = 100{,}000{,}000$ dollars squared, so $\sigma = \$10{,}000$; $\mu = \$390{,}000$.
The desired interval is: $\mu \pm 2\sigma = \$390{,}000 \pm 2(\$10{,}000) = (\$370{,}000, \$410{,}000)$.

3-74 Bullets: $CV = \frac{\sigma}{\mu}(100) = \frac{18(100)}{224} = 8.04\%$

Trailblazers: $CV = \frac{\sigma}{\mu}(100) = \frac{12(100)}{195} = 6.15\%$

The Bullets have the greater relative dispersion.

3-76 Regular MBA: $\bar{x} = 24.8, s = 2.486, CV = (s/\bar{x})(100) = \frac{2.486(100)}{24.8} = 10.02\%$

Evening MBA: $\bar{x} = 30.4, s = 2.875, CV = (s/\bar{x})(100) = \frac{2.875(100)}{30.4} = 9.46\%$

There is not much difference between the two groups.

3-78 Company 1: $CV = \frac{\sigma}{\mu}(100) = \frac{5.3(100)}{28} = 18.93\%$

Company 2: $CV = \frac{\sigma}{\mu}(100) = \frac{4.8(100)}{37.8} = 12.70\%$

Company 1 pursued the riskier strategy.

3-80

Employee	John	Jeff	Mary	Tammy
$\bar{x}$	66.67	67.40	71.83	61.20
s	3.78	1.14	9.70	4.66
$CV=100(\sigma/\bar{x})$	5.67%	1.69%	13.50%	7.61%

Jeff is the best employee, since he has the lowest coefficient of variation.

3-82 Configuration 1: $CV = \frac{\sigma}{\mu}(100) = \frac{4.8(100)}{34.8} = 13.79\%$

Configuration 2: $\quad CV = \frac{\sigma}{\mu}(100) = \frac{7.5(100)}{25.5} = 29.41\%$

Configuration 3: $\quad CV = \frac{\sigma}{\mu}(100) = \frac{3.8(100)}{37.5} = 10.13\%$

The third configuration has the least relative variation.

3-84 This statement is incorrect, because it completely ignores the variability in yards gained per carry. If the Raiders gain 85 yards once in every 200 carries, but gain 1.382 yards in the other 199 carries, then they average 3.6 yards a carry, but will rarely get a first down.

3-86 Since military manpower levels, staffing, and salaries are known with a fair amount of certainty, we could probably assign Curve A to the distribution of outcomes of actual officer salaries. Food purchases are also known with a fair amount of certainty, but prices are less certain. Thus, the distribution of outcomes of actual food purchases would probably fit Curve B. Finally, Curve C would be the best fit for the distribution of outcomes of actual aircraft maintenance since this expense would be fairly uncertain.

3-88 Perhaps the company has changed its hiring policy; specifically, it may now be hiring some less experienced sales reps and providing them with on-the-job training. This action, however, would also reduce the mean; thus hiring less experienced sales reps could not change the variation without also changing the mean. Perhaps the company is simultaneously hiring some highly experienced sales reps along with the less experienced ones. The higher scores of the former would balance out the lower scores of the latter, leaving the mean relatively unchanged while the variation increased.

3-90 The later period will show both a higher mean and a higher variability.

3-92 a)

x	17	21	44	50	79	86	140	178	203
x^2	289	441	1936	2500	6241	7396	19600	31684	41209

$\sum x = 818 \qquad \sum x^2 = 111{,}296$

$$\mu = \frac{\sum x}{N} = \frac{818}{9} = 90.8889$$

b) The median is the (9+1)/2 = 5th element, that is, 79.

c) Since each observation appears only once, there is no mode.

d) Since the data tend to cluster at the lower end of the range and the mean is pulled up by the the three high observations, the median is the better measure of central tendency.

e) $\sigma^2 = \sum x^2/\text{N} - \mu^2 = 111,296/9 - 90.8889^2 = 4105.43 \qquad \sigma = \sqrt{4105.43} = 64.074$

3-94 Formula #1: $\quad CV = \frac{\sigma}{\mu}(100) = \frac{35(100)}{700} = 5.0\%$

Formula #2: $\quad CV = \frac{\sigma}{\mu}(100) = \frac{16(100)}{300} = 5.3\%$

Formula #2 is relatively less accurate because it has the greater coefficient of variation.

3-96 a) Neither the variance nor the standard deviation would be a good measure of the variability.

b) The data clearly came from two distinct populations: one for weekdays (Sunday through Thursday), the other for weekends (Friday and Saturday). This fact would be lost if we lumped all seven days together to compute a single measure of variability.

3-98 To find the deciles, we arrange the 40 observations in ascending order:

	9.1	10.4	11.9	12.8	13.7	14.6	15.8	16.9	17.8	18.8
	9.3	10.6	12.1	13.0	13.9	14.7	16.0	17.1	18.0	19.0
	9.6	10.9	12.4	13.3	14.2	15.0	16.3	17.4	18.3	19.3
	9.9	11.2	12.7	13.6	14.5	15.3	16.6	17.7	18.6	19.6
deciles	1st	2nd	3rd	4th	5th	6th	7th	8th	9th	10th

Eighty percent of the trucks delivered fewer than 17.8 tons.

3-100 a) Range will give the least information because it considers only the highest and lowest observations. Standard deviation considers all of the data.

b) The range is certainly the easier to compute of the two measures.

c) In this case, since the data are fairly evenly spread out between the low of 210 days and the high of 231 days, the range is a reasonable measure of the variability in the data, so it may not be necessary to consider one of the other measures.

3-102 a) Listing the data in ascending order:

0 2 4 4 5 6 7 8 10 11 14 19 21 29

we see that the median is 7.5 days, halfway between the 7th and 8th observations.

b) $\mu = \sum x/N = 140/14 = 10$ days

3-104 The weekly news magazines would probably have the highest average readerships, the medical journals the smallest average readerships, with the monthly magazines somewhere in the middle.

Monthly magazines and medical journals, with many low circulation items and few high circulation items are likely to be skewed to the right. There are only a few weekly news magazines, so it's difficult to assess the skewness of this distribution.

3-106 a) Listing the data in ascending order:

4.77 4.89 4.91 5.02 5.05 5.22 5.24 5.27
5.75 5.99 6.01 6.02 6.05 6.11 6.11 6.11

we see that the median is 5.51 mpg, halfway between the 8th and 9th observations.

b) $x = \sum x/n = 88.52/16 = 5.5325$ mpg

c)

Class (mpg)	4.77-5.03	5.04-5.30	5.31-5.57	5.58-5.84	5.85-6.11
Frequency	4	4	0	1	7

The modal class is 5.85-6.11 mpg.

d) It depends. If she is ordering fuel for only one car, she should be cautious and use use the modal value. If she is ordering fuel for several cars running in the same race, the mean or median is probably ok.

3-108 a) Listing the data in ascending order:

21 23 25 26 26 32 33 33 33 34 36 37 37 37 37 43 45 47 47 56

we see that the median is 35 bulbs, halfway between the 10th and 11th observations.

$\bar{x} = \sum x/n = 708/20 = 35.4$ bulbs

b) Since the mean is bigger than the median, the distribution is skewed to the right.

3-110 a)

Class (mm)	Frequency	Cumulative Frequency
$\leq$ 1.00	12	12
1.01-1.50	129	141
1.51-2.00	186	327
2.01-2.50	275	602
2.51-3.00	341	943
3.01-3.50	422	1365
3.51-4.00	6287	7652
4.01-4.50	8163	15815
4.51-5.00	6212	22027
5.01-5.50	2416	24443
$\geq$5.51	1019	25462

The modal class is 4.01-4.50mm, which is also the median class. The median value is approximately

$$\tilde{m} = \left(\frac{(n+1)/2 - (F+1)}{f_m}\right) w + L_m = \left(\frac{25463/2 - 7653}{8163}\right)(.5) + 4.01 = 4.32\text{mm}$$

b) The coarse 3.5 mm screen suffices to remove at least half of the debris.

3-112 $\bar{x} = (7.41 + 7.24 + 5.15 + 5.09 + 4.61 + 2.77 + 2.67 + 2.00 + 0.14)/9 = \$4.12$

median = 5th observation = \$4.61

The median is better, since the mean is distorted by the observation for Southwest (\$0.14), which is clearly an outlier.

CHAPTER 4

PROBABILITY I: INTRODUCTORY IDEAS

4-2 The FDA conducted extensive experiments, exposing some laboratory animals to saccharin and keeping other animals (the "control group") free of saccharin intake. These experiments indicated that, with all other factors held constant (as much as possible), animals exposed to saccharin were more likely to develop cancer than those not exposed. This is an example of sampling from a larger population (the population of all laboratory animals, or, in a larger sense, the entire animal kingdom).

The warning statement is very much a probability statement for at least two reasons. First, it addresses the likelihood of cancer in laboratory animals under two sets of conditions, with the conclusion that the likelihood of cancer is higher under conditions including exposure to saccharin. Second, it claims that humans and laboratory animals are likely to react similarly to saccharin intake. Clearly, the FDA deems this likelihood to be high enough to merit the use of the warning statement.

4-4 This decision involves estimates of consumer preference, brand loyalty, competitor response, and numerous other factors, which all involve uncertainty. Thus the only estimates possible are probability-based.

4-6 a) (ball, strike) (ball, ball) (strike, strike) (strike, ball)

b) (ball, ball, ball) (strike, ball, ball)
(ball, ball, strike) (strike, ball, strike)
(ball, strike, strike) (strike, strike, ball)
(ball, strike, ball) (strike, strike, strike)

4-8

Sum	Ways to get this sum (card, die)	Probability
0	impossible	0
2	(1, 1)	1/54
3	(1, 2), (2, 1)	2/54
8	(2, 6), (3, 5), ..., (7, 1)	6/54
9	(3, 6), (4, 5), ..., (8, 1)	6/54
12	(6, 6), (7, 5), ..., (10, 2)	5/54
14	(8, 6), (9, 5), (10, 4)	3/54
16	(10, 6)	1/54

4-10 a) The segments are collectively exhaustive, because they are considered to be the only ones worthy of special campaigns. They are not mutually exclusive, because more than one campaign can be funded, depending on the total amount spent.

b) The list of collectively exhaustive and mutually exclusive events for the spending decision (with A = minorities, B = business people, etc.) is:

A only	A,C only	B,E only	A,C,E only
B only	A,D only	C,D only	A,D,E only
C only	A,E only	C,E only	C,D,E only
D only	B,C only	D,E only	
E only	B,D only		

c) Yes. The new list, which would again be both collectively exhaustive and mutually exclusive, is:

B,C only	A,C,D only
B,E only	A,D,E only

4-12

a) 6/26

b) 5/26

c) 1/2

d) 1/4

4-14 If we let P(50 – 74%) = P, then we know that:

P(75 - 99%) = 1/2 P and P(25 - 49%) = 2/5 P

Further, P(0 - 24%) = 0 and P(100%) = 1/20 = .05

Thus, 2/5 P + P + 1/2 P = 1.9P = .95, which means that P = .5

Therefore,
P(0 - 24%) = 0
P(25 - 49%) = .20
P(50 - 74%) = .50
P(75 - 99%) = .25
P(100%) = .05

4-16

a) subjective

b) relative frequency or subjective

c) classical

d) relative frequency

e) classical

f) relative frequency or subjective

4-18 P(A) = 21/100 P(B) = 29/100 P(C) = 38/100

P(A or B) = 45/100 P(A or C) = 50/100 P(B but not (Aor C)) = 20/100

4-20

a) Comparing the two expressions shows that when A and B are mutually exclusive, P(A and B) = 0. This is in line with the definition of mutually exclusive, which states that two events are mutually exclusive if they cannot both occur at the same time.

b) The correct expression would be:

$$P(\text{A or B or C}) = P(A) + P(B) + P(C) - P(\text{A and B}) - P(\text{A and C}) - P(\text{B and C}) + P(\text{A, B, and C})$$

c) If A and B are mutually exclusive, then by definition P(A and B) = 0. Also, if A and B cannot occur simultaneously, then A, B, and C cannot occur simultaneously. Therefore, P(A, B, and C) = 0. The correct expression becomes:

P(A or B or C) = P(A) + P(B) + P(C) – P(A and C) – P(B and C)

d) If A and C are also mutually exclusive, then P(A and C) = 0. Therefore, the expression would be:

P(A or B or C) = P(A) + P(B) + P(C) – P(B and C)

e) Finally, if all the events are mutually exclusive of each other, then

P(A or B or C) = P(A) + P(B) + P(C)

4-22 D = disk-drive failure K = keyboard failure

a) P(D or K) = P(D) + P(K) – P(D and K)
= P(D) + 3P(D) – .05 = 4P(D) – .05 = .20

Thus, 4P(D) = .25, so P(D) = .0625

b) $P(K) = 2P(D) = .125$
Thus, $P(D \text{ or } K) = P(D) + P(K) - .05 = .0625 + .125 - .05 = .1375$
This means that the computer is 86.25% resistant to disk-drive and/or keyboard failure.

4-24 a) $P(\text{Boy}_2 \mid \text{Girl}_1) = 1/2$ b) $P(\text{Girl}_2 \mid \text{Girl}_1) = 1/2$

4-26 a) 6/32 b) 6/32 c) 1/32

4-28 a) $P(\text{A passes} \mid \text{B fails}) = P(\text{A passes}) = .02$

b) $P(\text{B passes} \mid \text{A passes}) = P(\text{B passes}) = .07$

c) $P(\text{A and B pass}) = P(\text{A passes})P(\text{B passes}) = (.02)(.07) = .0014$

4-30 a) $P(\text{GL, DH, and DC approve}) = (.85)(.80)(.82) = .5576$

b) $P(\text{GL and DH approve, DC doesn't approve}) = (.85)(.80)(.18) = .1224$

4-32 a) $P(1, 2, 3, 4) = (.75)(.82)(.87)(.9) = .481545$

b) $P(1, \text{not } 2, \text{not } 3, 4) = (.75)(.18)(.13)(.9) = .015795$

c)
$$\begin{aligned} P(\text{one noticed}) &= P(1, \text{not } 2, \text{not } 3, \text{not } 4) + P(\text{not } 1, 2, \text{not } 3, \text{not } 4) \\ &\quad + P(\text{not } 1, \text{not } 2, 3, \text{not } 4) + P(\text{not } 1, \text{not } 2, \text{not } 3, 4) \\ &= (.75)(.18)(.13)(.1) + (.25)(.82)(.13)(.1) \\ &\quad + (.25)(.18)(.87)(.1) + (.25)(.18)(.13)(.9) \\ &= .01360 \end{aligned}$$

d) $P(\text{not } 1, \text{not } 2, \text{not } 3, \text{not } 4) = (.25)(.18)(.13)(.1) = .000585$

e) $P(\text{not } 3, \text{not } 4) = (.13)(.1) = .013$

4-34
$$P(A \mid C) = \frac{P(\text{A and C})}{P(C)} = \frac{1/7}{1/3} = \frac{3}{7}$$

$$P(C \mid A) = \frac{P(\text{A and C})}{P(A)} = \frac{1/7}{3/14} = \frac{2}{3}$$

$P(\text{B and C}) = P(B \mid C)P(C) = (5/21)(1/3) = 5/63$

$$P(C \mid B) = \frac{P(\text{B and C})}{P(B)} = \frac{5/63}{1/6} = \frac{10}{21}$$

4-36
$$P(\text{alcoholic} \mid \text{male}) = \frac{P(\text{male and alcoholic})}{P(\text{male})} = \frac{.21}{.59} = .356$$

4-38 H = hurricane in eastern Gulf, F = hurricane hits Florida

a) $P(\text{H and F}) = P(F \mid H)P(H) = (.76)(.85) = .646$

b) $P(F \mid H) = (3/4)(.76) = .57$
$P(\text{H and F}) = P(F \mid H)P(H) = (.57)(.85) = .4845$ [Note: $.4845 = (3/4)(.646)$]

4-40 P(upgrade and favorable evaluation)
$= P(\text{favorable evaluation})\ P(\text{upgrade} \mid \text{favorable evaluation}) = .65(.85) = .5525$

4-42 A = pilots strike D = drivers strike

a) $P(\text{A and D}) = P(A \mid D)P(D) = (.90)(.65) = .585$

b) $P(D \mid A) = P(\text{A and D})/P(A) = .585/.75 = .78$

4-44 R = received payment

Event	P(Event)	P(R | Event)	P(R and Event)	P(Event | R)
Personal Call	.7	.75	.525	.525/.71 = .739 (a)
Phone Call	.2	.60	.120	.120/.71 = .169 (b)
Letter	.1	.65	.065	.065/.71 = .092 (c)
			P(R) = .710	

4-46

Event	P(Event)	P(Storm | Event)	P(Storm & Event)	P(Event | Storm)
Dry	0.20	0.30	0.06	0.06/0.61 = 0.0984
Moist	0.45	0 .60	0.27	0.27/0.61 = 0.4426
Wet	0.35	0 .80	0.28	0.28/0.61 = 0.4590
			P(Storm) = 0.61	

The probability of a thunderstorm is 0.61. The probability of moist conditions, given that the picnic was cancelled (i.e., given that there was a thunderstorm) is 0.4426.

4-48 G = play on grass T = play on turf I = incur knee injury
P(I | T) = (1.5)P(I | G) = .42 ⇒ P(I | G) = .42/1.5 = .28

a) P(I) = P(I and T) + P(I and G) = (.42)(.4) + (.28)(.6) = .336

b) P(G | I) = P(I and G)/P(I) = (.28)(.6)/.336 = .5

4-50 L_{10}/G_{10} = lose/gain more than 10 seats
L_6/G_6 = lose/gain 6 - 10 seats
S =lose or gain 5 or fewer sets
R_2/F_2 = unemployment rises/falls by 2 percent or more
U = unemployment changes by less than 2 percent

a)

Event	P(Event)	P(G_6 | Event)	P(G_6 and Event)	P(Event | G_6)
R_2	.25	.15	.0375	.0375/.3150
U	.45	.35	.1575	.1575/.3150
F_2	.30	.40	.1200	.1200/.2150 = .3810
			P(G_6) = .3150	

b)

Event	P(Event)	P(S | Event)	P(S and Event)	P(Event | S)
R_2	.25	.15	.0375	.0375/.1350
U	.45	.15	.0675	.0675/.1350 = .5000
F_2	.30	.10	.0300	.0300/.1350
			P(S) = .1350	
			P(RL) = .2097	

4-52 The difference in rates would seem to suggest that the risk or probability of dying is greater as one gets older (common sense tells us that). Thus, this type of protection costs more because the company will have a greater probability of having to pay the claim before it has had the chance to collect very much in premiums. On the other hand, the higher rates for young drivers suggests that young drivers have a greater probability of having an accident, making it necessary for them to pay more for this protection.

4-54 Using past data on the rate of restaurant failures in an area or in the nation, econometricians calculate the relative frequency of restaurant failures and use this as an estimate of the chances of any given restaurant failing in this time period.

4-56 a) 1/5 b) 2/5 c) 3/5 d) classical

4-58 a) No. The item "he is nominated for vice-president" cannot occur during midterm elections.

b) No. He can, for example, win his party's re-election nomination and be re-elected.
Yes. Winning and losing his party's re-election nomination, for example are mutually exclusive.

c) No. Another possible event is "he is not re-elected."

4-60 a) No. For example, the firm might have won 5 other bigger contracts.

b) No. The cousin need not be the uncle's child.

c) Yes.

d) No. The promotion may come before the discovery of embezzlement.

4-62 a) $P(\text{plant will be a polluter}) = \dfrac{\text{observed \# of polluting plants}}{\text{total \# of observations}}$

$= \frac{2}{6} = .3333$

b) This probability was determined using the relative frequency of occurrence method.

c) The calculated probability is an inaccurate estimate of the true probability. The number of observations is extremely small, and the observations were made for plants which are not exactly like the proposed plant. Therefore, the probability estimate actually takes on many of the characteristics of a subjective estimate.

4-64 E = Engulf and Devour take over
R = R. A. Venns takes over

a) $P(E) = 7/20 = .35$ $\quad P(R) = 6/15 = .4$ $\quad P(E \text{ and } R) = 0$
$P(E \text{ or } R) = P(E) + P(R) - P(E \text{ and } R) = .35 + .4 - 0 = .75$

b) There is not sufficient information to answer the question.

4-66 A0/B0 = Flight 100/200 on time
A5/B5 = Flight 100/200 5 minutes late/early
A10/B10 = Flight 100/200 10 minutes late/early

a) $P(\text{collision}) = P(A5 \text{ and } B5) + P(A0 \text{ and } B10) + P(A10 \text{ and } B0)$
$= (.03)(.02) + (.95)(.01) + (.02)(.97)$
$= .0295 > .025$; diversion required

b) $P(\text{collision} \mid A5) = P(B5 \mid A5) = P(B5) = .02 < .025$; no diversion required

c) $P(\text{collision} \mid B5) = P(A5 \mid B5) = P(A5) = .03 > .025$; diversion required

4-68 a) $P(\text{Family}) = \frac{120}{300} = 0.4$

b) $P(\text{Spouse or other relative}) = \frac{6 + 15}{300} = \frac{21}{300} = 0.07$

4-70 B = passenger bumped
K = passenger takes Kansas City flight
A = passenger takes Atlanta flight
D = passenger takes Detroit flight

Event	P(Event)	P(B | Event)	P(B and Event)	P(Event | B)
A	.55	.07	.0385	.0385/.0670 = .5746 (a)
K	.20	.08	.0160	.0160/.0670 = .2388 (b)
D	.25	.05	.0125	.0125/.0670 = .1866 (c)
			P(S) = .0670	

4-72 a)

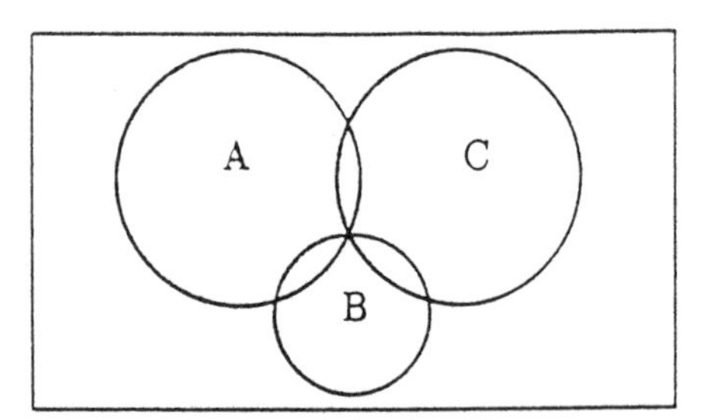

b)

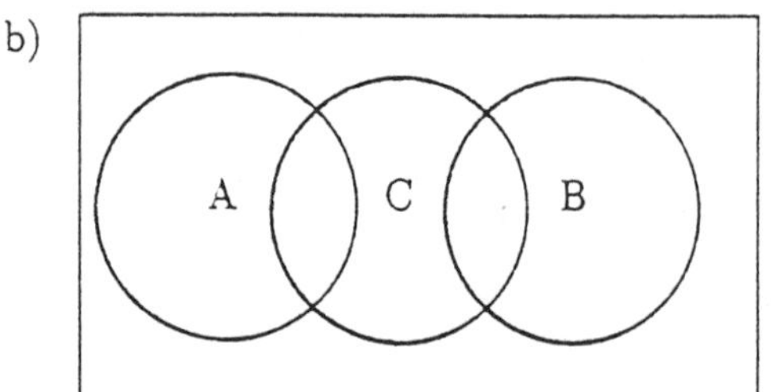

c)

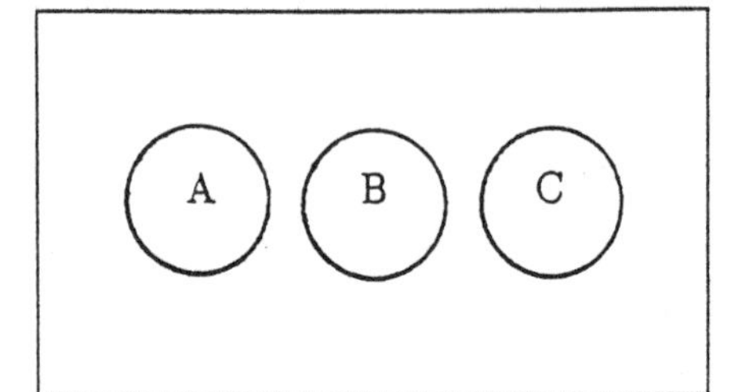

d)

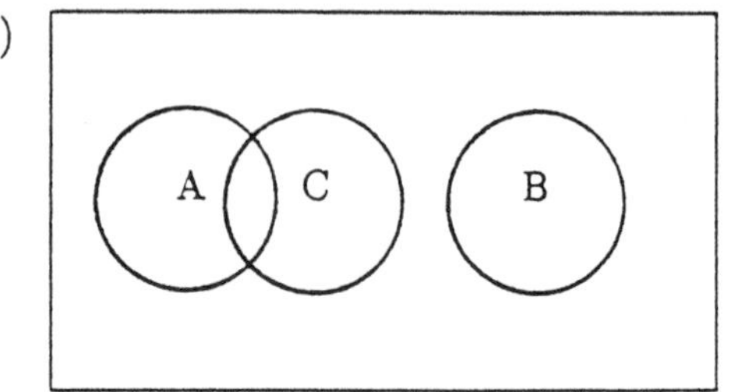

4-74 a) E1 = engine 1 fails   E2 = engine 2 fails
P(E1 and E2) = P(E1)P(E2 | E1) = .05(.10) = .005

b) B = recalled for brakes   S = steering flaw
Since the two events are statistically independent,
P(B and S) = P(B)P(S) = (.15)(.02) = .003

c) F = files on return   C = cheats on return
P(C and F) = P(C | F)P(F) = (.25)(.70) = .175

4-76 a) P(neutral | Chapel Hill) = 3/15 = .2
P(strongly opposed | Chapel Hill) = 2/15 = .1333

b) P(strongly supports) = (6 + 3 + 1)/45 = 10/45 = .222

c) P(neutral or slightly opposed | Raleigh or Lumberton) = (4 + 3 + 3 + 5)/30
= 15/30 = .5

4-78 a) P(damaged) = 565/10000 = .0565
P(overripe) = 1135/10000 = .1135

b) P(Ecuador or Honduras) = 1

c) P(Honduras | overripe) = 295/1135 = .2599

d) P(damaged or overripe) = P(damaged) + P(overripe) − P(damaged and overripe)
If damaged and overripe are mutually exclusive, then
P(damaged and overripe) = 0, so
P(damaged or overripe) = .0565 + .1135 = .1700
However, if they are not exclusive, then we can't find P(damaged or overripe) since we don't know P(damaged and overripe).

4-80 a) P(8-ball first) = 1/15 = .0667

b) P(8-ball first, second or third) = 3/15 = .2000

c) P(8-ball last) = 1/15 = .0667

4-82 a) $P(\text{death} \mid \text{vaccine}) = .02(.03) + .98(.0005) = .0011$
$P(\text{death} \mid \text{no vaccine}) = .3(.04) = .0120$

b) $P(\text{death}) = P(\text{death} \mid \text{vaccine})\,P(\text{vaccine}) + P(\text{death} \mid \text{no vaccine})\,P(\text{no vaccine})$
$= .0011(.25) + .0120(.75) = .0093$

4-84

Event	P(Event)	P(4 tears | Event)	P(4 tears & Event)
Normal	.4	$(.07)^4 = .00002401$	.000009604
Fast	.6	$(.14)^4 = .00038416$	.000230496
		P(4 tears) =	.000240100

Thus, P(Fast | 4 tears) = .000230496/.00024010 = .96

4-86 a) $6/50,000 + 6/50,000 - (6/50,000)^2 = 0.00024$, or about once in 4167 six-hour flights.

b) $6/50,000$

c) $(6/50,000)^2$

CHAPTER 5

PROBABILITY II: DISTRIBUTIONS

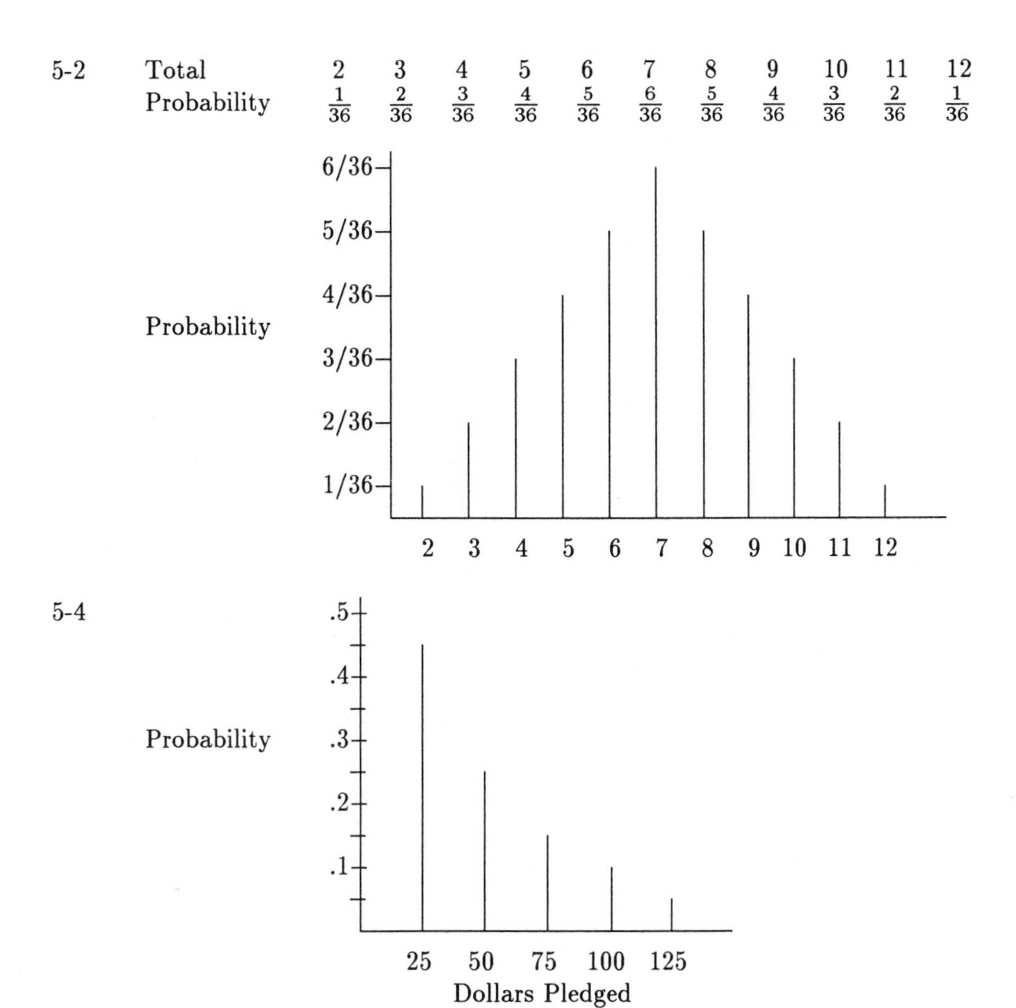

5-2

Total	2	3	4	5	6	7	8	9	10	11	12
Probability	$\frac{1}{36}$	$\frac{2}{36}$	$\frac{3}{36}$	$\frac{4}{36}$	$\frac{5}{36}$	$\frac{6}{36}$	$\frac{5}{36}$	$\frac{4}{36}$	$\frac{3}{36}$	$\frac{2}{36}$	$\frac{1}{36}$

5-6

# of Jets Sold	Probability
2500	.2
5000	.7
5500	.1

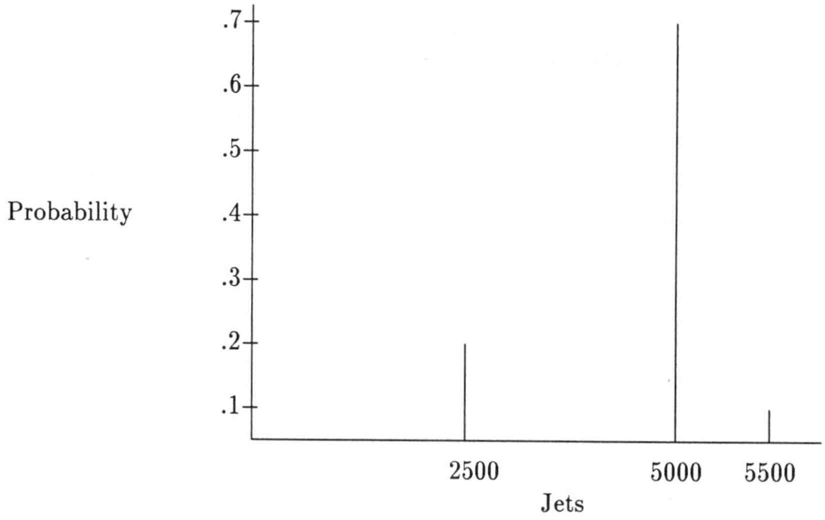

Probability

.7
.6
.5
.4
.3
.2
.1

2500 5000 5500

Jets

5-8 a)

Outcome (1)	P(Outcome) (2)	b) (1) × (2)
\$ 8,000	.05	400
9,000	.15	1350
10,000	.25	2500
11,000	.30	3300
12,000	.20	2400
13,000	.05	650
	1.00	\$10,600 = expected value

5-10 The expected cost is

$$0(0.35) + 50(0.25) + 100(0.15) + 150(0.10) + 200(0.08) + 250(0.05) + 300(0.02) = \$77$$

so Jim should not pay \$100 for the warranty.

5-12

Months to Settle (1)	Probability (2)	(1) × (2)	Months to Settle (1)	Probability (2)	(1) × (2)
1	.02	.01	11	.05	.55
2	.02	.04	12	.06	.72
3	.01	.03	13	.07	.91
4	.02	.08	14	.08	1.12
5	.02	.10	15	.10	1.50
6	.03	.18	16	.09	1.44
7	.04	.28	17	.08	1.36
8	.03	.24	18	.08	1.44
9	.04	.36	19	.07	1.33
10	.04	.40	20	.06	1.20
				1.00	13.29

The expected number of months to settle = 13.29.

5-14 Truck Division:

# of Letters Lost (1)	Frequency (2)	Probability (3)	(1) × (3)
0	1	.0833	.0000
1	2	.1667	.1667
2	2	.1667	.3333
3	2	.1667	.5000
4	2	.1667	.6667
5	2	.1667	.8333
6	0	.0000	.0000
7	1	.0833	.5833
	12	1.0000	3.0833

Air Division:

# of Letters Lost (1)	Frequency (2)	Probability (3)	(1) × (3)
0	2	.1667	.0000
1	1	.0833	.0833
2	2	.1667	.3333
3	1	.0833	.2500
4	3	.2500	1.0000
5	1	.0833	.4167
6	1	.0833	.5000
7	1	.0833	.5833
	12	1.0000	3.1667

He investigates the air division.

5-16 Cost = 6300/2 = \$3150 per car
Revenue = 35 − 2.50 = \$32.50 per car per day
Revenue = 32.50 × 312 = \$10,140 per year
10140 − 3150 = \$6990 = opportunity cost per car

Loss Table

States of Nature

	13	14	15	16	17	18	
Probability	.08	.15	.22	.25	.21	.09	
Action							Expected Loss
13	0	6690	13980	20970	27960	34950	18383.70
14	3150	0	6990	13980	20970	27960	12204.90
15	6300	3150	0	6990	13980	20970	7547.10
16	9450	6300	3150	0	6990	13980	5120.10 ←
17	12600	9450	6300	3150	0	6990	5228.10
18	15750	12600	9450	6300	3150	0	7465.50

The optimal number of cars is 16.

5-18 binomial ($n = 7,\ p = .2$)

a) $P(r=5) = \left(\frac{7!}{5!2!}\right)(.2)^5(.8)^2 = .0043$

b) $P(r > 2) = 1 - P(r \leq 2) = 1 - [P(r=0) + P(r=1) + P(r=2)]$

$= 1 - \left(\frac{7!}{0!7!}\right)(.2)^0(.8)^7 - \left(\frac{7!}{1!6!}\right)(.2)^1(.8)^6 - \left(\frac{7!}{2!5!}\right)(.2)^2(.8)^5$

$= 1 - .2097 - .3670 - .2753 = .1480$

c) $P(r < 8) = 1$

d) $P(r \geq 4) = P(r > 2) - P(r = 3) = .1480 - \left(\frac{7!}{3!4!}\right)(.2)^3(.8)^4$

$= .1480 - .1147 = .0333$

5-20

	n	p	$\mu = np$	$\sigma = \sqrt{npq}$
a)	15	.20	3.00	1.549
b)	8	.42	3.36	1.396
c)	72	.26	4.32	2.015
d)	29	.49	14.21	2.692
e)	642	.21	134.82	10.320

5-22 a) P(more than 2 flaws) = 1 − P(0 flaws) − P(1 flaw) − P(2 flaws)

$= 1 - \left(\frac{10!}{0!10!}\right)(.02)^0(.98)^{10} - \left(\frac{10!}{1!9!}\right)(.02)^1(.98)^9 - \left(\frac{10!}{2!8!}\right)(.02)^2(.98)^8$

$= 1 - .8171 - .1667 - .0153 = .0009$

b) $P(0 \text{ flaws}) = \left(\frac{10!}{0!10!}\right)(.02)^0(.98)^{10} = .8171$

5-24 binomial, $n = 15$, p to be determined

a) claim 1: $P(r \geq 4) = .9095 \Rightarrow p = .40$
claim 2: $np > 7 \Rightarrow p > 7/15 = .467$
The claims are not consistent.

b) No. The assistants claim that $p = .40$

5-26 Let r be a binomial random variable, with $n = 15$ and $p = 0.3$, representing the number of browsing customers who buy something.

a) $P(r \geq 1) = .9953$

b) $P(r \geq 4) = .7031$

c) $P(r = 0) = 1 - P(r \geq 1) = 1 - .9953 = .0047$

d) $P(r \leq 4) = 1 - P(r \geq 5) = 1 - .4845 = .5155$

5-28 $\lambda = 4$, $e^{-4} = .0183$

a) $P(x = 0) = \frac{e^{-4}(4)^0}{0!} = .0183$

b) $P(x = 2) = \frac{e^{-4}(4)^2}{2!} = .1465$

c) $P(x = 4) = \frac{e^{-4}(4)^4}{4!} = .1954$

d) $P(x \geq 5) = 1 - [P(x = 0) + P(x = 1) + P(x = 2) + P(x = 3) + P(x = 4)]$

$$= 1 - \left[\frac{e^{-4}(4)^0}{0!} + \frac{e^{-4}(4)^1}{1!} + \frac{e^{-4}(4)^2}{2!} + \frac{e^{-4}(4)^3}{3!} + \frac{e^{-4}(4)^4}{4!}\right]$$

$$= 1 - [.0183 + .0733 + .1465 + .1954 + .1954] = .3711$$

5-30 Using Appendix Table 4(b),

a) $P(x \leq 3) = P(x=0) + P(x=1) + P(x=2) + P(x=3)$
$= .0022 + .0137 + .0417 + .0848 = .1424$

b) $P(x \geq 2) = 1 - P(x=0) - P(x=1) = 1 - .0022 - .0137 = .9841$

c) $P(x=6) = .1605$

d) $P(1 \leq x \leq 4) = P(x=1) + P(x=2) + P(x=3) + P(x=4)$
$= .0137 + .0417 + .0848 + .1294 = .2696$

5-32 $P(x=0) + P(x=1) + P(x=2) + P(x=3) \quad = e^{-4.1}\left[1 + \frac{4.1}{1!} + \frac{(4.1)^2}{2!} + \frac{(4.1)^3}{3!}\right]$

$= .01657(1 + 4.1 + 8.405 + 11.4868) = .4141$

So the probability of 4 or more stops per hour is 1 − .4141 = .5859, and so Ford should move the employee.

5-34 The mean number of malfunctions per 50 calculators is 50(.04) = 2, so

a) $P(\geq 3 \text{ malfunctions}) \quad = 1 \; - P(0 \text{ malfunctions}) - P(1 \text{ malfunction}) - P(2 \text{ malfunctions})$

$= 1 - \frac{e^{-2}(2)^0}{0!} - \frac{e^{-2}(2)^1}{1!} - \frac{e^{-2}(2)^0}{2!}$

$= 1 - .135335 - .270671 - .270671 = .323323$

b) $P(0 \text{ malfunctions}) = .135335$

5-36 binomial, $n = 1000$, $p = .005$; $\lambda = np = 5$, $e^{-5} = .00674$

a) $P(0) = \frac{e^{-5}(5)^0}{0!} = .00674$

b) $P(10) = \frac{e^{-5}(5)^{10}}{10!} = .01813$

c) $P(15) = \frac{e^{-5}(5)^{15}}{15!} = .00016$

5-38 $\mu = np = 50(.25) = 12.5$, $\sigma = \sqrt{npq} = \sqrt{50(.25)(.75)} = 3.062$

a) $P(r > 10) = P\left(z > \frac{10.5 - 12.5}{3.062}\right) = P(z > -.65) = .5 + .2422 = .7422$

b) $P(r < 18) = P\left(z < \frac{17.5 - 12.5}{3.062}\right) = P(z < 1.63) = .5 + .4484 = .9484$

c) $P(r > 21) = P\left(z > \frac{21.5 - 12.5}{3.062}\right) = P(z > 2.94) = .5 - .4984 = .0016$

d) $P(9 < r < 14) = P\left(\frac{9.5 - 12.5}{3.062} < z < \frac{13.5 - 12.5}{3.062}\right)$

$= P(-.98 < z < .33) = .3365 + .1293 = .4658$

5-40 a) $\mu = 35(.15) = 5.25$, $\sigma = \sqrt{35(.15)(.85)} = 2.112$

$P(7 \leq r \leq 10) = P\left(\frac{6.5 - 5.25}{2.112} \leq z \leq \frac{10.5 - 5.25}{2.112}\right)$

$= P(0.59 \leq z \leq 2.49) = .4936 - .2224 = .2712$

b) $\mu = 29(.25) = 7.25, \quad \sigma = \sqrt{29(.25)(.75)} = 2.332$

$$P(r \geq 9) = P\left(z \geq \frac{8.5 - 7.25}{2.332}\right) = P(z \geq 0.54) = .5 - .2054 = .2946$$

c) $\mu = 84(.42) = 35.28, \quad \sigma = \sqrt{84(.42)(.58)} = 4.524$

$$P(r \leq 40) = P\left(z \leq \frac{40.5 - 35.28}{4.524}\right) = P(z \leq 1.15) = .5 + .3749 = .8749$$

d) $\mu = 63(.11) = 6.93, \quad \sigma = \sqrt{63(.11)(.89)} = 2.483$

$$P(r \geq 10) = P\left(z \geq \frac{9.5 - 6.93}{2.483}\right) = P(z \geq 1.04) = .5 - .3508 = .1492$$

e) $\mu = 18(.67) = 12.06, \quad \sigma = \sqrt{18(.67)(.33)} = 1.995$

$$\begin{aligned} P(9 \leq r \leq 12) &= P\left(\frac{8.5 - 12.06}{1.995} \leq z \leq \frac{12.5 - 12.06}{1.995}\right) \\ &= P(-1.78 \leq z \leq 0.22) = .4625 + .0871 = .5496 \end{aligned}$$

5-42 $\mu = 5.07, \sigma = .07$

$$\begin{aligned} P(5.05 \leq x \leq 5.15) &= P\left(\frac{5.05 - 5.07}{.07} \leq z \leq \frac{5.15 - 5.07}{.07}\right) \\ &= P(-.29 \leq z \leq 1.14) = .1141 + .3729 = .4870 \end{aligned}$$

5-44 $\mu = 4.02, \sigma = .08$ (since 68% of the probability in a normal distribution falls in the interval $\mu \pm \sigma$)

$$\begin{aligned} P(3.9 < x < 4.1) &= P\left(\frac{3.9 - 4.02}{.08} < z < \frac{4.1 - 4.2}{.08}\right) \\ &= P(-1.5 < z < 1) = .4332 + .3413 = .7745 < .8 \end{aligned}$$

The prototype does not satisfy the medical standards.

5-46 $\mu = 44, \sigma = 12$

a) $$\begin{aligned} P(33 \leq x \leq 42) &= P\left(\frac{33 - 44}{12} \leq z \leq \frac{42 - 44}{12}\right) \\ &= P(-.92 \leq z \leq -.17) = .3212 - .0675 = .2537 \end{aligned}$$

b) $$P(x < 30) = P\left(z < \frac{30 - 44}{12}\right) = P(z < -1.17) = .5 - .3790 = .1210$$

c) $$\begin{aligned} P(x < 25 \text{ or } x > 60) &= P\left(z < \frac{25 - 44}{12}\right) + P\left(z > \frac{60 - 44}{12}\right) \\ &= P(z < -1.58) + P(z > 1.33) \\ &= (.5 - .4429) + (.5 - .4082) = .1489 \end{aligned}$$

5-48 a) $$P(x > 72000) = P\left(z \geq \frac{72000 - 67000}{4000}\right) = P(z \geq 1.25) = .5 - .3944 = .1056$$

b) The expected loss to future business is \$5000(.1056) = \$528. Since this can be avoided if he hires the new employees for \$200, he should hire them

5-50 a) $\mu = 325, \sigma = 60$, given Estimate I is accurate

$$P(x > 350 \mid EI) = P\left(z > \frac{350 - 325}{60}\right) = P(z > .42) = .5 - .1628 = .3372$$

b) $\mu = 300, \sigma = 50$, given Estimate II is accurate

$$P(x > 350 \mid EII) = P\left(z > \frac{350 - 300}{50}\right) = P(z > 1) = .5 - .3413 = .1587$$

c,d) Let X denote $x > 350$.

Event	P(Event)	P(X | Event)	P(X and Event)	P(Event | X)
EI	.5	.3372	.16860	.16860/.24795 = .68 (c)
EII	.5	.1587	.07935	.07935/.24795 = .32 (d)
			P(X) = .24795	

5-52 a) normal b) Poisson c) binomial d) normal

5-54 A random variable is considered to be discrete if it can assume only a limited number of values. In other words, all the values could be listed. A continuous random variable can assume any value within a given range. It is impossible, therefore, to list all the possible values of a continuous random variable. Often a discrete random variable can assume so many values that we assume it is continuous for ease of calculation.

Just as there are discrete and continuous random variables, so also are there discrete and continuous probability functions, each associated with the appropriate type of random variable.

5-56 Since $n = 200$ and $p = .03$, we can approximate the binomial distribution for the number of checks that bounce by a Poisson distribution with $\lambda = np = 200(.03) = 6$.

a) $P(r = 10) = \frac{6^{10} e^{-6}}{10!} = .0413$

b) $P(r = 5) = \frac{6^{5} e^{-6}}{5!} = .1606$

5-58

Number of Cases Researched (1)	Frequency (2)	Probability (3)	(1) × (3)
4	3	.3	1.2
5	3	.3	1.5
6	2	.2	1.2
7	1	.1	.7
8	1	.1	.8
		Expected number of cases =	5.4

5.4 × 2 interns per case = 10.8, or 11, interns to hire. (But if each case requires two interns for the whole summer, then perhaps 12 should be hired.)

5-60 a) normal b) binomial c) Poisson d) normal

5-62 Since $n = 200$ and $p = .02$, we can approximate the binomial distribution for the number of flights more than 10 minutes early or late by a Poisson distribution with $\lambda = np = 200(.02) = 4$.

$$P(r = 0) = \frac{4^{0} e^{-4}}{0!} = e^{-4} = .01832$$

$$P(r = 10) = \frac{4^{10} e^{-4}}{10!} = \frac{1048576(.018316)}{3628800} = .00529$$

5-64 Let r be a binomial random variable with $n = 15$ and $p = .25$, representing the number of applicants who were rejected.

a) $P(r = 4) = .2252$

b) $P(r = 8) = .0131$

c) $P(r < 3) = .0134 + .0668 + .1559 = .2361$

d) $P(r > 5) = 1 - .0134 - .0668 - .1559 - .2552 - .2552 - .1651 = .0884$

5-66 Let r be a binomial random variable with $n = 15$ and $p = .12$, representing the number who repeat the course.

a) $P(r < 6) = 1 - .0047 - .0008 - .0001 = .9944$

b) $P(r = 5) = .0208$

c) $P(r \leq 2) = .1470 + .3006 + .2870 = .7346$

5-68 a)

Year	1992	1993	1994	1995	1996
Number of Copies	15,000	8,500	10,000	7,700	11,000
Probability	.10	.25	.45	.10	.10

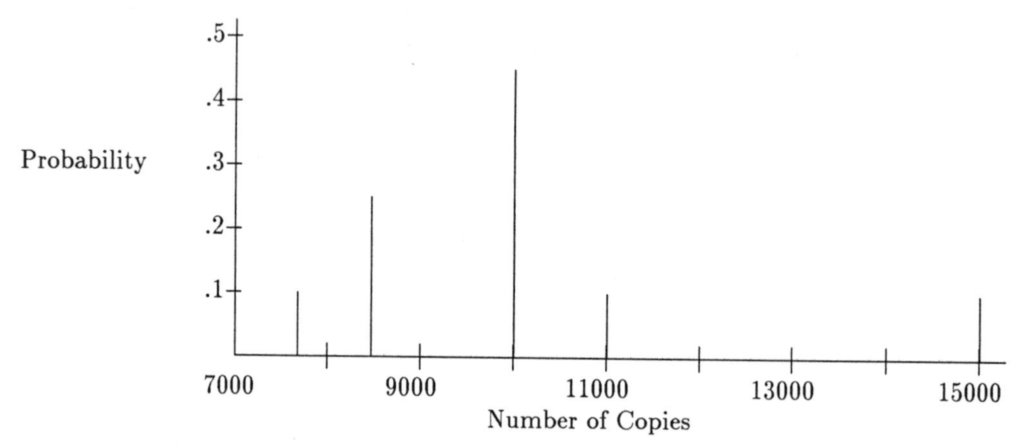

b) She should order the maximum number of pamphlets which could be demanded in 1989, i.e., 15,000.

5-70

# of Coats Sold (1)	Probability (2)	(a) Profit (3)	(a) (2) × (3)	(b) Profit (4)	(b) (2) × (3)
8	.10	1160	116	920	92
10	.20	1240	248	1080	216
12	.25	1320	330	1240	310
14	.45	1400	630	1400	630
	1.00		1324		1248

a) Heidi's expected profit is $1324.

b) Heidi's expected profit will fall to $1248.

5-72 a) Yes: knowing that the average number of customers varies from day to day enables the manager to know how inconvenient the ARM maintenance will be, on average.

b) Yes: only the Friday data are relevant if the maintenance is rescheduled for Friday.

5-74 Let x be a normal random variable with $\mu = 16{,}050$ and $\sigma = 2500$, representing the number of fans attending.

a) $P(x > 20{,}000) = P\left(z > \frac{20000 - 16050}{2500}\right) = P(z > 1.58) = .5 - .4429 = .0571$

b) $P(x < 10{,}000) = P\left(z < \frac{10000 - 16050}{2500}\right) = P(z < -2.42) = .5 - .4922 = .0078$

c) $P(14{,}000 < x < 17{,}500) = P\left(\frac{14000 - 16050}{2500} < z < \frac{17500 - 16050}{2500}\right)$

$= P(-.82 < z < .58) = .2939 + .2190 = .5129$

5-76 Let x be a normal random variable with $\mu = 24$ and $\sigma = 7.5$, representing the length of a lease.

a) $P(x \geq 28) = P\left(z \geq \frac{28 - 24}{7.5}\right) = P(z \geq .53) = .5 - .2019 = .2981$

b) $P(x < 12) = P\left(z < \frac{12 - 24}{7.5}\right) = P(z < -1.60) = .5 - .4452 = .0548$

5-78 Since we are not given the number of graduates from each school, we will treat each school as one observation. Thus we are computing statistics and probabilities per school, rather than per graduate.

a,c)

Post-MBA Salaries		Job Offers
x	x^2	x
89.93	8087.4049	3.02
84.64	7163.9296	2.96
83.21	6923.9041	2.92
100.80	10160.6400	3.47
102.63	10532.9169	3.60
67.82	4599.5524	2.68
58.52	3424.5904	2.45
100.48	10096.2304	2.43
74.01	5477.4801	2.74
80.50	6480.2500	3.25
70.49	4968.8401	2.78
74.28	5517.5184	2.69
95.41	9103.0681	2.40
69.89	4884.6121	2.69
71.97	5179.6809	2.40
70.66	4992.8356	2.12
61.89	3830.3721	2.58
69.88	4883.2144	3.09
71.97	5179.6809	2.34
54.72	2994.2784	2.19
1553.70	124480.9998	54.80

For the post-MBA salaries:

$\mu = \sum x/N = 1553.70/20 = 77.6850$

$\sigma^2 = \sum x^2/N - \mu^2$
$= 124480.9998/20 - 77.6850^2 = 189.0908$

$\sigma = \sqrt{\sigma^2} = \sqrt{189.0908} = 13.7510$

b) 1. $P(x > 100) = P(z > (100 - 77.685)/13.751) = P(z > 1.62) = .5 - .4474 = .0526$

2. $P(x \leq 60) = P(z \leq (60 - 77.685)/13.751) = P(z \leq -1.29) = .5 + .4015 = .0985$

3. $P(75 \leq x \leq 95) = P((75 - 77.685)/13.751 \leq z \leq (95 - 77.685)/13.751)$
$= P(-0.20 \leq z \leq 1.26) = .0793 + .3962 = .4755$

c) $\lambda = \sum x/N = 54.8/20 = 2.74$

d) 1. $P(x < 2) = P(x{=}0) + P(x{=}1)$

$$= e^{-2.74}\left(\frac{(2.74)^0}{0!} + \frac{(2.74)^1)}{1!}\right) = .06457(1 + 2.74) = .2415$$

2. $P(x{=}2 \text{ or } x{=}3) = P(x{=}2) + P(x{=}3)$

$$= e^{-2.74}\left(\frac{(2.74)^2}{2!} + \frac{(2.74)^3)}{3!}\right) = .06457(3.7538 + 3.4285) = .4638$$

3. $P(x > 3) = 1 - P(x \leq 3) = 1 - .2415 - .4638 = .2947$

5-80 $\mu = np = 15(.12) = 1.8$, $\sigma = \sqrt{npq} = \sqrt{15(.12)(.88)} = 1.259$

a) $P(r < 6) = P\left(z < \frac{5.5 - 1.8}{1.259}\right) = P(z < 2.94) = .5 + .4984 = .9984$
(exact answer = .9943; 0.4% error)

b) $P(r = 5) = P\left(\frac{4.5 - 1.8}{1.259} < z < \frac{5.5 - 1.8}{1.259}\right) = P(2.14 < z < 2.94)$
$= .4984 - .4838 = .0146$ (exact answer = .0208; 30% error)

c) $P(r < 3) = P\left(z < \frac{2.5 - 1.8}{1.259}\right) = P(z < .56) = .5 + .2123 = .7123$
(exact answer = .7346; 3% error)

In contrast to problem 79, where $np = 5.28 > 5$ and $nq = 6.72 > 5$ and the largest percentage error was only 1%, here where $np < 5$, we encounter larger errors.

5-82 Opportunity loss = 0.70 - 0.35 = \$0.35 per rose = \$350 per thousand roses
Obsolescence loss = 0.35 - 0.10 = \$0.25 per rose = \$250 per thousand roses
Actions and states of nature are given in thousands of roses.

Loss Table

	States of Nature				
	15	20	25	30	
Probability	.10	.30	.40	.20	
Action					Expected loss
15	0	1750	3500	5250	2975
20	1250	0	1750	3500	1525
25	2500	1250	0	1750	975 ←
30	3750	2500	1250	0	1625

The optimal order of roses to produce is 25 thousand.

5-84 $n = 4$, $p = .11$

$$P(r = 0) = \frac{4!}{0!4!}(.11)^0(.89)^4 = .62742$$

$$P(r = 1) = \frac{4!}{1!3!}(.11)^1(.89)^3 = .31019$$

$$P(r = 2) = \frac{4!}{2!2!}(.11)^2(.89)^2 = .05751$$

$$P(r = 3) = \frac{4!}{3!1!}(.11)^3(.89)^1 = .00474$$

$$P(r = 4) = \frac{4!}{4!0!}(.11)^4(.89)^0 = .00015$$

5-86 a) Assuming that approvals are independent from loan to loan, and that all loans have the same .8 probability of approval, then

$$\mu = np = 1460(.8) = 1168, \quad \sigma = \sqrt{npq} = \sqrt{1460(.8)(.2)} = 15.28$$

b) $\mu = 1327(.77) = 1021.79$, $\sigma = \sqrt{1327(.77)(.23)} = 15.33$

5-88
$$P(r \le 102 \mid n = 300, p = .4) = P\left(z \le \frac{102.5 - 300(.4)}{\sqrt{300(.4)(.6)}}\right)$$
$$= P(z \le -2.06) = .5 - .4803 = .0197$$

This small probability makes it unlikely that the 40% estimate is correct. The estimate appears to be overly optimistic.

5-90 a) $p = \text{P(pay decrease)} = 10/39 = .2564$

a.1) $P(x = 5) = \frac{6!}{5!1!}(.2564)^5(.7436)^1 = .0049$

a.2) If at least five got raises, then at most one saw a drecrease:

$$P(x \leq 1) = \frac{6!}{0!6!}(.2564)^0(.7436)^6 + \frac{6!}{1!5!}(.2564)^1(.7436)^5$$
$$= .1691 + .3498 = .5189$$

a.3) If fewer than four got raises, then at least three saw decreases:

$$P(x \geq 3) = 1 - P(x \leq 2) = 1 - .5189 - \frac{6!}{2!4!}(.2564)^2(.7436)^4$$
$$= 1 - .5189 - .3015 = .1796$$

b) Using a calculator, $\overline{x} = 10.977$, $s = 20.912$

c.1) $P(x \geq 25) = P\left(z \geq \frac{25 - 10.977}{20.912}\right) = P(z \geq 0.67) = .5 - .2486 = .2514$

c.2) $P(x \leq 5) = P\left(z \leq \frac{5 - 10.977}{20.912}\right) = P(z \leq -0.29) = .5 - .1141 = .3859$

c.3) $P(-15 \leq x \leq 15) = P\left(\frac{-15 - 10.977}{20.912} \leq z \leq \frac{15 - 10.977}{20.912}\right) = P(-1.24 \leq z \leq 0.19)$

$$= .3925 + .0753 = .4678$$

CHAPTER 6

SAMPLING AND SAMPLING DISTRIBUTIONS

6-2 They are not necessarily mutually exclusive. For example, a situation could arise in which the history (or lack of history) of the population of interest indicates no clear-cut trends. Thus, the personal opinion of the individual conducting the study could be that probability sampling is best.

6-4 Probability samples involve more rational analysis and planning at the beginning of a study and usually take more time and money than judgment samples.

6-6 From what we have been told in the problem, Jean's position is apparently quite defensible. Perhaps what makes statistical sampling unique is that it permits statistical inference to be made about a population and its parameters. This is apparently what Jean has done. There are no hard and fast rules as to the size of the sample that must be drawn before inferences can be made. Specifically there is nothing magic about the 50% mark. Common sense would seem to point out that gathering data from 50% of some populations might tend to be almost as difficult as gathering data from the entire population--for instance, the population of the United States, or the world. The defense for Jean's position lies in empirical evidence and some explanation and reasoning with the project leaders, educating them about the abilities of statistical inference.

6-8 In (b), the distributions have greater between-group variance and less within-group variance than in (a).

6-10 Assuming a non-leap year:

1/6	1/24	2/11	3/1	3/19	4/6	4/24	5/12	5/30	6/17
7/5	7/23	8/10	8/28	9/15	10/3	10/21	11/8	11/26	12/14

6-12 The probability that a 4, 7, or 2 will appear is .10, since each digit is equally likely to be selected by a random number generator. Since there are 115 ten digit numbers, we would expect to see each number appear 11.5 times. Actually, the numbers appear as follows:

Number	4	7	2
Appearances	16	13	10

Assuming that our table is random, we would expect, in a larger version of this table, the number of appearances of any digit in any given position to be close to one tenth of the possible appearances. The large deviations from 12.5 in this case are due to insufficient sample size.

6-14 No. Between noon and 5 pm on a weekday, no one will be at home if both parents work, and their children are of pre-school age. Thus, the heaviest users of daycare centers are excluded from a poll on daycare centers.

6-16 Every seventh is better. If every fifth lot were inspected, then the output of only two machines would be inspected. If every seventh lot is inspected, then each machine's output will be inspected.

6-18 We need to remember that when we use stratified sampling, we are only acknowledging that the population is already divided into groups of different sizes. However, these groups are required to be relatively homogeneous before we can use this sampling method. In the case under question, Mary proposes dividing the population into groups such as urban, suburban, and rural; examination of the differences among these groups should convince you that they are reasonably homogeneous, and thus stratified sampling would work.

6-20 Sampling error, the chance error which is due to the particular elements selected as a sample of the population.

6-22 A sample mean overestimating the true mean is no better than one under- estimating the true mean. In this case, 30 cents is closer than 35 cents to the true mean, in absolute differences:

$$|\ 30 - 31.4\ | = 31.4 - 30 = 1.4 < |\ 35 - 31.4\ | = 35 - 31.4 = 3.6$$

6-24 Average weekly sales have decreased from 3538 cartons to 3462 cartons.

6-26 The information gathered concerns mean customer satisfaction for groups of 30 customers, not for single customers, so it is a sample from the sampling distribution of the mean of samples of size 30 drawn from the customer population. It is not a sample from the customer population.

6-28 a) $n = 19 \qquad \mu = 18 \qquad \sigma = 4.8 \qquad \sigma_{\bar{x}} = \sigma/\sqrt{n} = 4.8/\sqrt{19} = 1.101$

$$P(16 < \bar{x} < 20) = P\left(\frac{16 - 18}{1.101} < z < \frac{20 - 18}{1.101}\right)$$
$$= P(-1.82 < z < 1.82) = 2(.4656) = .9312$$

b) The same as (a), since the distribution is continuous

c) $n = 48, \qquad \mu = 18 \qquad \sigma = 4.8 \qquad \sigma_{\bar{x}} = \sigma/\sqrt{n} = 4.8/\sqrt{48} = 0.693$

$$P(16 < \bar{x} < 20) = P\left(\frac{16 - 18}{0.693} < z < \frac{20 - 18}{0.693}\right)$$
$$= P(-2.89 < z < 2.89) = 2(.4981) = .9962$$

6-30 $\mu = 375 \qquad \sigma = 48 \qquad P(-1.96 < z < 1.96) = .95$

Want $P(370 < \bar{x} < 380) \geq .95 \Rightarrow P\left(0 < z < \frac{380 - 375}{48/\sqrt{n}}\right) \geq .475 \Rightarrow \frac{5\sqrt{n}}{48} \geq 1.96$

$$\Rightarrow \sqrt{n} \geq 18.816 \Rightarrow n \geq 354.04$$

A sample size of 355 will be sufficient.

6-32 $\mu = \$62{,}000 \qquad \sigma = \$4{,}200$

a) $n = 1 \qquad \sigma_{\bar{x}} = \sigma/\sqrt{n} = \$4{,}200$

$$P(\bar{x} \geq 65{,}000) = P\left(\frac{\bar{x} - \mu}{\sigma_{\bar{x}}} \geq \frac{65{,}000 - 62{,}000}{4{,}200}\right) = P(z \geq .71)$$
$$= .5 - .2611 = .2389$$

b) The probability is less. More sampling decreases the standard error of the sampling distribution of the mean.

$n = 2 \qquad \sigma_{\bar{x}} = \$4200/\sqrt{2} = \$2970$

$$P(\bar{x} \geq 65000) = P\left(\frac{\bar{x} - \mu}{\sigma_{\bar{x}}} \geq \frac{65000 - 62000}{2970}\right) = P(z \geq 1.01)$$
$$= .5 - .3438 = .1562$$

The decrease is .0827, or 35%.

6-34 a) $\mu = 4300$ $\sigma = 730$ $n = 3$ $\sigma_{\bar{x}} = \sigma/\sqrt{n} = 730/\sqrt{3} = 421.5$

For a set of monitors to last 13000 hours, they must each last 4333.33 hours on average.

$$P(\bar{x} \geq 4333.33) = P\left(\frac{\bar{x} - \mu}{\sigma_{\bar{x}}} \geq \frac{4333.33 - 4300}{421.5}\right)$$

$$= P(z > .08) = .5 - .0319 = .4681$$

b) For the set to last at most 12630 hours, the average life cannot exceed 4210 hours.

$$P(\bar{x} \leq 4210) = P\left(\frac{\bar{x} - \mu}{\sigma_{\bar{x}}} \leq \frac{4210 - 4300}{421.5}\right) = P(z \leq -.21) = .5 - .0832 = .4168$$

6-36 The sample size of 48 is large enough to use the central limit theorem.

$\mu = 110$ $\sigma = 64$ $n = 48$ $\sigma_{\bar{x}} = \sigma/\sqrt{n} = 64/\sqrt{48} = 9.238$

$$P(\bar{x} < 120) = P\left(\frac{\bar{x} - \mu}{\sigma_{\bar{x}}} < \frac{120 - 110}{9.238}\right) = P(z < 1.08) = .5 + .3599 = .8599 > .80$$

The overhaul will not be ordered.

6-38 $\mu = 120$ $\sigma = 12$ $n = 60$

a) $\mu_{\bar{x}} = \mu = 120$

b) $\sigma_{\bar{x}} = \sigma/\sqrt{n} = 12/\sqrt{60} = 1.549$

c) $P(\bar{x} > 123.8) = P\left(\frac{\bar{x} - \mu}{\sigma_{\bar{x}}} > \frac{123.8 - 120}{1.549}\right) = P(z > 2.45) = .5 - .4929 = .0071$

d) $$P(117 < \bar{x} < 122) = P\left(\frac{117 - 120}{1.549} < \frac{\bar{x} - \mu}{\sigma_{\bar{x}}} < \frac{122 - 120}{1.549}\right)$$

$$= P(-1.94 < z < 1.29) = .4738 + .4015 = .8753$$

6-40 $N = 75$ $n = 32$ $\mu = 364$ $\sigma = \sqrt{18} = 4.243$

a) $\sigma_{\bar{x}} = \frac{\sigma}{\sqrt{n}}\sqrt{\frac{N - n}{N - 1}} = \frac{4.243}{\sqrt{32}}\sqrt{\frac{75-32}{75-1}} = 0.572$

b) $$P(363 \leq \bar{x} \leq 366) = P\left(\frac{363-364}{0.572} \leq z \leq \frac{366-364}{0.572}\right)$$

$$= P(-1.75 \leq z \leq 3.50) = .4599 + .5 = .9599$$

c) With replacement, $\sigma_{\bar{x}} = \sigma/\sqrt{n} = 4.243/\sqrt{32} = 0.750$

6-42 $N = 80$ $\mu = 8.2$ $\sigma = 2.1$ $\sigma_{\bar{x}} = \frac{\sigma}{\sqrt{n}}\sqrt{\frac{N - n}{N - 1}}$

a) $n = 16$ $\sigma_{\bar{x}} = \frac{2.1}{4}\sqrt{\frac{64}{79}} = .4725$

b) $n = 25$ $\sigma_{\bar{x}} = \frac{2.1}{5}\sqrt{\frac{55}{79}} = .3504$

c) $n = 49$ $\sigma_{\bar{x}} = \frac{2.1}{7}\sqrt{\frac{31}{79}} = .1879$

6-44 $N = 45$ $n = 9$ $\mu = 225{,}000$ $\sigma = 39{,}000$

$$\sigma_{\bar{x}} = \frac{\sigma}{\sqrt{n}}\sqrt{\frac{N - n}{N - 1}} = \frac{39{,}000}{3}\sqrt{\frac{36}{44}} = 11{,}759$$

$$P(9\overline{x} \geq 2{,}100{,}000) = P\left(\overline{x} \geq \frac{2{,}100{,}000}{9} = 233{,}333.33\right)$$

$$= P\left(\frac{\overline{x} - \mu}{\sigma_{\overline{x}}} \geq \frac{233{,}333.33 - 225{,}000}{11{,}759}\right)$$

$$= P(z \geq .71) = .5 - .2611 = .2389$$

6-46 With 250(.36) = 90 contributors, the average donation must be between \$1222.22 and \$1333.33 for the total to be between \$112,950 and \$116,100. Each 4% gift has $\mu = .04(32{,}000) = \$1{,}280$ and $\sigma = .04(9{,}600) = \$384$.

$$\sigma_{\overline{x}} = \frac{\sigma}{\sqrt{n}}\sqrt{\frac{N-n}{N-1}} = \frac{384}{\sqrt{90}}\sqrt{\frac{160}{249}} = 32.4467$$

$$P(1222.22 < \overline{x} < 1333.33) = P\left(\frac{1222.22 - 1280}{32.4467} < \frac{\overline{x} - \mu}{\sigma_{\overline{x}}} < \frac{1333.33 - 1280}{32.4467}\right)$$

$$= P(-1.78 < z < 1.64) = .4625 + .4495 = .9120$$

6-48 $N = 145$ $\quad n = 36$ $\quad \sigma = 1200$ $\quad \sigma_{\overline{x}} = \frac{\sigma}{\sqrt{n}}\sqrt{\frac{N-n}{N-1}} = \frac{1200}{\sqrt{36}}\sqrt{\frac{109}{144}} = 174.01$

$$P(\mu - 200 \leq \overline{x} < \mu + 200) = P\left(\frac{-200}{174.01} \leq \frac{\overline{x} - \mu}{\sigma_{\overline{x}}} \leq \frac{200}{174.01}\right)$$

$$= P(-1.15 \leq z \leq 1.15) = .3749 + .3749 = .7498$$

6-50 Judgmental, because the sample (those skates inspected) is determined by whether the skate is Crash's size.

6-52 Stratified sampling is used when we have reason to believe that we can divide the population into groups which are relatively homogeneous. In this instance, the manager feels that the residents fall into various age and income levels. Since he is trying to measure residents' attitudes, stratified sampling may be his best method, especially since he feels residents in different age and income groups may have different attitudes.

6-54 $\mu = 26$ $\quad \sigma = 5.65$ $\quad \sigma_{\overline{x}} = 5.65/\sqrt{n}$

We want to find n such that: $P(25 \leq \overline{x} \leq 27) \geq .9544 = P(-2 \leq z \leq 2)$

$$\Rightarrow P\left(\frac{25 - 26}{5.65/\sqrt{n}} \leq \frac{\overline{x} - \mu}{\sigma_{\overline{x}}} \leq \frac{27 - 26}{5.65/\sqrt{n}}\right) \geq P(-2 \leq z \leq 2)$$

$$\Rightarrow P(-.177\sqrt{n} \leq z \leq .177\sqrt{n}) \geq P(-2 \leq z \leq 2)$$

$$\Rightarrow .177\sqrt{n} \geq 2 \Rightarrow \sqrt{n} \geq 11.3 \Rightarrow n \geq 127.7$$

Thus, a sample of at least 128 customers is needed.

6-56 In this situation, Fargo Lanna is not constrained by (1) cost, (2) time, (3) destruction of population members, or (4) accessibility of the population. Nevertheless, sampling would be entirely appropriate for this company's purpose. It will be able to obtain the same information without expending nearly the effort required to poll all employees. The company may get the same job done by using a sample, and will be able to reassign the clerical staff earlier.

6-58 Again Simmons is incorrect. A sampling distribution of means is a frequency distribution of the means of all possible samples. It is not in any sense a graph of the individual observations in sample contributions.

6-60 a)

Job Title	Wage x	x^2	Job Title	Wage x	x^2
Assembler A	10.72	114.9184	Packaging-wrapping packer	9.04	81.7216
Assembler B	9.13	83.3569	Packer - heavy	10.08	101.6064
Assembler C	7.98	63.6804	Packer - light	8.82	77.7924
Carpenter, maintenance	13.58	184.4164	Painter - maintenance	12.72	161.7984
Chemical compounder	12.64	159.7696	Painter - spray	9.78	95.6484
Chemical mixer	11.19	125.2161	Plastic injection molder	9.72	94.4784
Degreaser operator	9.11	82.9921	Polisher & buffer A	10.24	104.8576
Drill press operator A	12.01	144.2401	Polisher & buffer B	9.59	91.9681
Drill press operator B	9.89	97.8121	Punch press die setter A	12.80	163.8400
Drill press operator C	9.51	90.4401	Punch press die setter B	11.31	127.9161
Electrician	15.37	236.2369	Punch press - heavy	9.75	95.0625
Grinding machine operator A	12.92	166.9264	Punch press - light	8.91	79.3881
Grinding machine operator B	9.89	97.8121	Punch press - setup/operate	11.32	128.1424
Group leader A	13.55	183.6025	Receiving clerk	9.98	99.6004
Group leader B	11.28	127.2384	Screw machine operator	16.01	256.3201
Guard-watchman	9.86	97.2196	Screw machine setup	12.40	153.7600
Inspector A	11.55	133.4025	Shear operator	10.45	109.2025
Inspector B	10.11	102.2121	Shipper/receiver	9.73	94.6729
Inspector C	8.57	73.4449	Shipping clerk	10.03	100.6009
Janitor - heavy	9.19	84.4561	Solderer A	5.69	32.3761
Janitor - light	8.26	68.2276	Solderer B	9.88	97.6144
Laborer-dock hand	9.26	85.7476	Stationary engineer	16.52	272.9104
Lathe operator-turret A	12.66	160.2756	Stock person A	9.71	94.2841
Lathe operator-turret B	10.62	112.7844	Stock person B	8.86	78.4996
Lift truck operator	10.52	110.6704	Test analyzer - junior	10.04	100.8016
Machine operator	9.82	96.4324	Test analyzer - senior	11.73	137.5929
Maintenance machinist A	15.31	234.3961	Tool & die maker A	17.66	311.8756
Maintenance machinist B	14.42	207.9364	Tool & die maker B	15.49	239.9401
Maintenance machinist C	12.07	145.6849	Tool & die maker C	11.72	137.3584
Maintenance person A	14.13	199.6569	Tool room machinist	13.55	183.6025
Maintenance person B	11.19	125.2161	Trucker, hand	9.00	81.0000
Modelmaker	16.35	267.3225	Warehouse person	9.87	97.4169
Numerical control A	13.99	195.7201	Welder - arc acetylene	12.69	161.0361
Numerical control B	10.69	114.2761	Welder - spot	10.01	100.2001
Packager	7.92	62.7264	Wirer A	10.67	113.8489
Packaging-wrapping operator	9.93	98.6049	Wirer B	8.81	77.6161
				799.77	9271.4231

$\mu = \sum x/N = 799.77/72 = \$11.108$

$\sigma^2 = \sum x^2/N - \mu^2 = 9271.4231/72 - 11.108^2 = 5.3821$

$\sigma = \sqrt{\sigma^2} = \sqrt{5.3821} = \$2.320$

b) $\overline{x} = \sum x/n = (7.98+15.37+15.37+9.86+10.52+16.35+12.80+10.45+10.67)/9$
$= \$12.152$

c) $\mu_{\overline{x}} = \mu = 11.108,\ \sigma_{\overline{x}} = \sigma/\sqrt{n} = 2.320/\sqrt{9} = \$0.773$

d) Probably not. The sample size is small (n=9), and the population, with 15 of its 72 members between \$9.50 and \$10.00 and a distinct skewness to the right, cannot be considered even approximately normal.

e) $P(10.5 \leq x \leq 11.7) = P((10.5 - 11.108)/0.773 \leq z \leq (11.7 - 11.108)/0.773)$
$= P(-0.79 \leq z \leq 0.77) = 0.2852 + 0.2794 = 0.5646$

6-62 $\mu = 41 \qquad \sigma = 8 \qquad n = 6 \qquad \sigma_{\overline{x}} = \sigma/n = 8/\sqrt{6} = 3.266$

$P(\overline{x} < 50) = P\left(\frac{\overline{x} - \mu}{\sigma_{\overline{x}}} < \frac{50 - 41}{3.266}\right) = P(z < 2.76) = .5 + .4971 = .9971$

6-64 Cost = Benefit

$$\Rightarrow 4n = \frac{5249}{\sigma_{\overline{x}}} = \frac{5249}{\frac{\sigma}{\sqrt{n}}} = \frac{5249\sqrt{n}}{265} = 19.81\sqrt{n} \Rightarrow \sqrt{n} = \frac{19.81}{4} = 4.95 \Rightarrow n = 24.5$$

Thus, she should sample at least 25 detectors.

6-66 a) Enumeration

b) Finite population

CHAPTER 7

ESTIMATION

7-2 a) Measuring an entire population may be impossible or infeasible, because of time and cost considerations, among other reasons.

b) A sample yields only an estimate and is subject to sampling errors.

7-4 a) An estimator is a sample statistic used to estimate a population parameter.

b) An estimate is a specific numerical value for an estimator, resulting from the particular sample which is observed.

7-6 It assures us that the estimator becomes more reliable with larger samples.

7-8 Using a calculator, $\bar{x} = 296.583$ people, $s = 40.751$ people

7-10 $n = 400$ $\quad x = 184$ $\quad \bar{p} = 184/400 = .46$

7-12 $\sigma = 1.4$ $\quad n = 60$ $\quad \bar{x} = 6.2$

a) $\sigma_{\bar{x}} = \sigma/\sqrt{n} = 1.4/\sqrt{60} = .181$

b) $\bar{x} \pm \sigma_{\bar{x}} = 6.2 \pm .181 = (6.019, 6.381)$

7-14 $\sigma = 0.8$ $\quad n = 421$ $\quad \bar{x} = 6.2$

a) $\sigma_{\bar{x}} = \sigma/\sqrt{n} = 0.8/\sqrt{421} = .0390$ pounds

b) $\bar{x} \pm 2\sigma_{\bar{x}} = 14.2 \pm 2(.039) = 14.2 \pm .078 = (14.122, 14.278)$

7-16 $\sigma = .9$ $\quad n = 75$ $\quad \bar{x} = 7$ $\quad \sigma_{\bar{x}} = \sigma/\sqrt{n} = .9/\sqrt{75} = .104$

$\bar{x} \pm 2\sigma_{\bar{x}} = 7 \pm 2(.104) = 7 \pm .208 = (6.79, 7.21)$

7-18 $\sigma = 8.3$ $\quad N = 621$ $\quad n = 76$ $\quad \bar{x} = 29.8$

$$\sigma_{\bar{x}} = \frac{\sigma}{\sqrt{n}}\sqrt{\frac{N-n}{N-1}} = \frac{8.3}{\sqrt{6}}\sqrt{\frac{621-76}{621-1}} = .893 \text{ students}$$

a) $\bar{x} \pm 2\sigma_{\bar{x}} = 29.8 \pm 2(.893) = 29.8 \pm 1.786 = (28.01, 31.59)$ students

b) We cannot be 95.5% certain that the average class size in Foresight County is less than that of Hindsight County. Therefore, we cannot conclude that Dee has met her goal.

7-20 The confidence interval is the range of an estimate, i.e., the interval between and including the upper confidence limit and the lower confidence limit.

7-22 a) High confidence levels produce wide intervals, so we sacrifice precision to gain confidence.

b) Narrow intervals result from low confidence levels, so we sacrifice confidence to gain precision.

7-24 No. It is based on the process of estimation and the expected results if the sampling process is repeated many times.

7-26 a) $\sigma = 5/2 \qquad x = 25$
$x \pm 1.96\sigma = 25 \pm 1.96(5/2) = 25 \pm 4.9 = (20.1, 29.9)$ minutes

b) $\sigma = 5/3 \qquad x = 15$
$x \pm 1.96\sigma = 15 \pm 1.96(5/3) = 15 \pm 3.267 = (11.73, 18.27)$ minutes

c) $\sigma = 5/5 \qquad x = 38$
$x \pm 1.96\sigma = 38 \pm 1.96(1) = (36.04, 39.96)$ minutes

d) $\sigma = 5/1 \qquad x = 20$
$x \pm 1.96\sigma = 20 \pm 1.96(5) = 20 \pm 9.8 = (10.2, 29.8)$ minutes

A confidence interval will contain the true population mean with the indicated percent confidence. A probability interval is one in which the value of the next observation will fall with the indicated probability.

7-28 $\sigma = 13.7 \qquad n = 250 \qquad \bar{x} = 112.4 \qquad \sigma_{\bar{x}} = \sigma/\sqrt{n} = 13.7/\sqrt{250} = 0.866$

a) $\bar{x} \pm 1.96\sigma_{\bar{x}} = 112.4 \pm 1.96(.866) = 112.4 \pm 1.697 = (110.70, 114.10)$

b) $\bar{x} \pm 2.58\sigma_{\bar{x}} = 112.4 \pm 2.58(.866) = 112.4 \pm 2.234 = (110.17, 114.63)$

7-30 $s = 1.2 \qquad n = 40 \qquad N = 700 \qquad \bar{x} = 4.3$

a) $\hat{\sigma}_{\bar{x}} = \frac{s}{\sqrt{n}}\sqrt{\frac{N-n}{N-1}} = \frac{1.2}{\sqrt{40}}\sqrt{\frac{700-40}{700-1}} = 0.184$

b) $\bar{x} \pm 1.64\hat{\sigma}_{\bar{x}} = 4.3 \pm 1.64(0.184) = 4.3 \pm 0.302 = (4.00, 4.60)$ typos per page.

7-32 $s = 3.2 \qquad n = 45 \qquad \bar{x} = 24.3 \qquad \hat{\sigma}_{\bar{x}} = s/\sqrt{n} = 3.2/\sqrt{45} = 0.477$

$\bar{x} \pm 1.96\hat{\sigma}_{\bar{x}} = 24.3 \pm 1.96(0.477) = 24.3 \pm .935 = (23.37, 25.24)$ minutes

7-34 $\hat{\sigma} = 41000 \qquad N = 12368 \qquad n = 750 \qquad \bar{x} = 250000 \qquad \frac{n}{N} = \frac{750}{12368} = .061 > .05$

$$\hat{\sigma}_{\bar{x}} = \frac{\hat{\sigma}}{\sqrt{n}}\sqrt{\frac{N-n}{N-1}} = \frac{41000}{\sqrt{750}}\sqrt{\frac{12368-750}{12368-1}} = 1451.1$$

$\bar{x} \pm 1.64\hat{\sigma}_{\bar{x}} = 250000 \pm 1.64(1451.1) = 250000 \pm 2380 = (\$247{,}620; \$252{,}380)$

7-36 $n = 55 \qquad \bar{p} = .1818$

a) $\hat{\sigma}_{\bar{p}} = \sqrt{\frac{\bar{p}\bar{q}}{n}} = \sqrt{\frac{.1818(.8182)}{55}} = .0520$

b) $\bar{p} \pm 1.645\hat{\sigma}_{\bar{p}} = .1818 \pm 1.645(.0520) = .1818 \pm .0855 = (.096, .267)$

7-38 $n = 200 \qquad \bar{p} = 174/200 = .87$

a) $\hat{\sigma}_{\bar{p}} = \sqrt{\frac{\bar{p}\bar{q}}{n}} = \sqrt{\frac{.87(.13)}{200}} = .0238$

b) $\bar{p} \pm 2.33\hat{\sigma}_{\bar{p}} = .87 \pm 2.33(.0238) = .87 \pm .0555 = (.815, .925)$

7-40 $N = 3000 \qquad n = 150 \qquad \frac{n}{N} = \frac{150}{3000} = .05 \qquad \bar{p} = .6$

$$\hat{\sigma}_{\bar{p}} = \sqrt{\frac{pq}{n}}\sqrt{\frac{N-n}{N-1}} = \sqrt{\frac{.6(.4)}{150}}\sqrt{\frac{3000-150}{3000-1}} = .0390$$

a) $\bar{p} \pm 1.96\hat{\sigma}_{\bar{p}} = .6 \pm 1.96(.0390) = .6 \pm .0764 = (.524, .676)$

b) $3000 \times (.524, .676) = (1572, 2028)$ accounts

7-42 $n = 45 \qquad \bar{p} = .6 \qquad \hat{\sigma}_{\bar{p}} = \sqrt{\frac{pq}{n}} = \sqrt{\frac{.6(.4)}{45}} = .0730$

$\bar{p} \pm 2.05\hat{\sigma}_{\bar{p}} = .6 \pm 2.05(.0730) = .6 \pm .1497 = (.450, .750)$

7-44 a) 1.761 b) 2.571 c) 2.878 d) 2.492 e) 3.250 f) 2.704

7-46 $s = 10 \qquad n = 12 \qquad \bar{x} = 62 \qquad \hat{\sigma}_{\bar{x}} = s/\sqrt{n} = 10/\sqrt{12} = 2.887$

$\bar{x} \pm t_{11,.025}\hat{\sigma}_{\bar{x}} = 62 \pm 2.201(2.887) = 62 \pm 6.354 = (55.65, 68.35)$

7-48 $s = 6.2 \qquad n = 21 \qquad \bar{x} = 72 \qquad \hat{\sigma}_{\bar{x}} = s/\sqrt{n} = 6.2/\sqrt{21} = 1.353$

$\bar{x} \pm t_{20,.01}\hat{\sigma}_{\bar{x}} = 72 \pm 2.528(1.353) = 72 \pm 3.420 = (68.58, 75.42)$

7-50 $s = 9 \qquad n = 9 \qquad \bar{x} = 31 \qquad \hat{\sigma}_{\bar{x}} = s/\sqrt{n} = 9/\sqrt{9} = 3$

$\bar{x} \pm t_{8,.05}\hat{\sigma}_{\bar{x}} = 31 \pm 1.860(3) = 31 \pm 5.58 = (25.42, 36.58)$ accidents

7-52 $p = .7 \qquad q = .3$

$.02 = 1.64\sqrt{\frac{pq}{n}} = 1.64\sqrt{\frac{.7(.3)}{n}}$, so $n = \left(\frac{1.64\sqrt{.7(.3)}}{.02}\right)^2 = 1412.04$, i.e., $n \geq 1413$

7-54 $p = .5 \qquad q = .5$

$.05 = 1.96\sqrt{\frac{pq}{n}} = 1.96\sqrt{\frac{.5(.5)}{n}}$, so $n = \left(\frac{1.96\sqrt{.5(.5)}}{.05}\right)^2 = 384.16$, i.e., $n \geq 385$

With $p = .75$ (or $p = .25$), $n = \left(\frac{1.96\sqrt{.25(.75)}}{.05}\right)^2 = 288.12$, i.e., $n \geq 289$

7-56 $.5 = 1.96\sigma/\sqrt{n} = 1.96(1.2)/\sqrt{n}$, so $n = \left(\frac{1.96(1.2)}{.5}\right)^2 = 22.1$, i.e., $n \geq 23$

7-58 $4 = 2.05\sigma/\sqrt{n} = 2.05(15)/\sqrt{n}$, so $n = \left(\frac{2.05(15)}{4}\right)^2 = 59.09$, i.e., $n \geq 60$

7-60 An interval estimate gives an indication of possible error through the extent of its range and through the probability associated with the interval. A point estimate is only a single number, and thus one needs additional information to determine its reliability.

7-62 Assume $p = q = .5$

$.01 = 1.96\sqrt{\frac{pq}{n}} = 1.96\sqrt{\frac{.5(.5)}{n}}$, so $n = \left(\frac{1.96(.5)}{.01}\right)^2 = 9604$

7-64 $s = .3$ $n = 57$ $\overline{x} = 23.2$

a) $\hat{\sigma} = s = .3$ mph

b) $\hat{\sigma}_{\overline{x}} = \hat{\sigma}/\sqrt{n} = .3/\sqrt{57} = .0397$ mph

c) $\overline{x} \pm 1.96\hat{\sigma}_{\overline{x}} = 23.2 \pm 1.96(.0397) = 23.2 \pm .0778 = (23.122, 23.278)$ mph

7-66 $\overline{x}$ is called the "best" estimator of the population mean because it exhibits all the qualities of good estimators we have discussed. It is unbiased, consistent, relatively efficient, and sufficient.

7-68 a) $2(.3944) = .7888$, i.e., 78.88%

b) $2(.4918) = .9836$, i.e., 98.36%

c) $2(.4535) = .9070$, i.e., 90.70%

7-70 $p = .85$

$.05 = 2.33\sqrt{\frac{pq}{n}} = 2.33\sqrt{\frac{.85(.15)}{n}}$, so $n = \left(\frac{2.33\sqrt{.85(.15)}}{.05}\right)^2 = 276.87$, i.e., $n \geq 277$

To be completely conservative, assume $p = .5$, so $n = \left(\frac{2.33\sqrt{.5(.5)}}{.05}\right)^2 = 542.89$, i.e., $n \geq 543$.

7-72 $n = 27$ $\overline{p} = 19/27 = 0.7037$

$\hat{\sigma}_{\overline{p}} = \sqrt{pq/n} = \sqrt{.7037(.2963)/27} = 0.791$

$\overline{p} \pm 1.96\hat{\sigma}_{\overline{p}} = 0.7037 \pm 1.96(0.0879)$

$= 0.7037 \pm 0.1723 = (0.5314,\ 0.8760)$

Since the entire interval is above 0.50, they can be more than 95% confident of breaking even at least half of the time. They should stay open on weeknights.

7-74 a) $s = 3.56\%$ (calculated in a LOTUS 1-2-3 spreadsheet)

b) $.5 = 2.58\sigma/\sqrt{n} = 2.58(3.56)/\sqrt{n}$, so $n = (2.58(3.56)/.5)^2 = 337.44$, i.e., $n \geq 338$.

7-76 $n = 19$, $\overline{x} = 2.88$, $s = 3.30$ (calculated in a LOTUS 1-2-3 spreadsheet)

$\hat{\sigma}_{\overline{x}} = s/\sqrt{n} = 3.30/\sqrt{19} = .7571$

$x \pm t_{18,.025}\hat{\sigma}_{\overline{x}} = 2.88 \pm 2.101(.7571) = 2.88 \pm 1.591 = (1.29\%,\ 4.47\%)$

Since $n < 30$, we must assume normality so that the t distribution can be used to form the confidence interval.

7-78 $\sigma = .2$ $n = 105$ $\overline{x} = 3.2$

a) $\sigma_{\overline{x}} = \sigma/\sqrt{n} = .2/\sqrt{105} = .0195$ apples

b) $\overline{x} \pm \sigma_{\overline{x}} = 3.2 \pm .0195 = (3.181, 3.219)$ apples

7-80 $s = 107.10$ $n = 200$ $\overline{x} = 425.39$

a) $\hat{\mu} = \overline{x} = \$425.39$
$\hat{\sigma} = s = \$107.10$

b) $\overline{x} \pm 1.96\hat{\sigma}_{\overline{x}} = .032 \pm 1.96\hat{\sigma}/\sqrt{n}$ $= 425.39 \pm 1.96(107.10)/\sqrt{200}$
$= 425.39 \pm 14.843 = (\$410.55, \$440.23)$

7-82 $\sigma = 1.4$ $n = 200$ $\overline{x} = 4.6$

a) $\sigma_{\overline{x}} = \sigma/\sqrt{n} = 1.4/\sqrt{200} = .0990$

b) $\mu \pm \sigma_{\overline{x}} = 5.2 \pm .099 = (5.10, 5.30)$

7-84 $n = 120$ $\overline{p} = .3333$ $\hat{\sigma}_{\overline{p}} = \sqrt{\frac{\overline{p}\overline{q}}{n}} = \sqrt{\frac{.3333(.6667)}{120}} = .0430$

$\overline{p} \pm 1.96\hat{\sigma}_{\overline{p}} = .3333 \pm 1.96(.0430) = .3333 \pm .0843 = (.249, .418)$

7-86 $s = .6$ $n = 186$ $\overline{x} = 66.3$

a) $\hat{\sigma}_{\overline{x}} = s/\sqrt{n} = .6/\sqrt{186} = .0440$ mph

b) $\overline{x} \pm 2\hat{\sigma}_{\overline{x}} = 66.3 \pm 2(.0440) = 66.3 \pm .088 = (66.212, 66.388)$ mph

c) Yes, since the entire interval lies below 67 mph

7-88 $6\sigma = 140 - 80 = 60$, so $\sigma = 10$
$5 = 1.64\sigma/\sqrt{n} = 1.64(10)/\sqrt{n}$, so $n = \left(\frac{1.64(10)}{5}\right)^2 = 10.75$, i.e., $n \geq 11$

CHAPTER 8

TESTING HYPOTHESES: ONE-SAMPLE TESTS

8-2 Theoretically, one could toss a coin a large number of times and see if the proportion of heads was very different from .5. Similarly by recording the outcomes of many dice rolls, one could see if the proportion of each side was very different from 1/6. You would need a large number of trials for each of these samples.

8-4 a) Assume hypothesis about population

b) Collect sample data

c) Calculate a sample statistic

d) Use sample statistic (c) to evaluate hypothesis (a)

8-6 We mean that we would not have reasonably expected to find that particular sample if in fact the hypothesis had been true.

8-8 $P(|z| \geq 1.75) = 2(.5 - .4599) = .0802$

8-10 $\sigma = 6000$ $n = 64$ $\bar{x} = 26100$ $\mu_{H_0} = 28500$

$$\mu \pm 2\sigma_{\bar{x}} = \mu \pm 2\sigma/\sqrt{n} = 28500 \pm 2(6000)/\sqrt{64}$$
$$= 28500 \pm 1500 = (27000, 30000)$$

Since $\bar{x} = 26100 < 27000$, Ned should not purchase the Stalwarts. Depending on how long ago Ned's last purchase was, it may no longer be reasonable to suppose that $\sigma = 6000$. If σ has increased sufficiently, Ned's decision could change.

8-12 $\sigma = 0.2$ $n = 42$ $\bar{x} = 2.2$ $\mu_{H_0} = 2.5$

$$\mu \pm 2.5\sigma_{\bar{x}} = \mu \pm 2.5\sigma/\sqrt{n} = 2.5 \pm 2.5(0.2)/\sqrt{42} = 2.5 + 0.077 = (2.423, 2.577)$$

Since $\bar{x} = 2.2 < 2.423$, it is unreasonable to see such sample results if μ really is 2.5 quarts; the store's claim is not correct.

8-14 A null hypothesis represents the hypothesis you are trying to reject. Alternative hypotheses represent all other possibilities.

8-16 Type I: the probability that we will reject the null hypothesis when in fact it is true.
Type II: the probability that we will accept the null hypothesis when in fact it is false.

8-18 The significance level of a test is the probability of a Type I error, i.e., the probability that we will reject the null hypothesis when it is, in fact, true. This is because it indicates the percentage of sample means that fall outside the limits of what we will accept as confirming the null hypothesis. Hence, if a sample mean falls outside these limits but is truly from the hypothesized population, it will lead to a Type I error.

8-20 a) t with 34 df (so we use the normal table) b) normal c) normal

d) t with 28 df e) t with 23 df

8-22 A two-tailed test of a hypothesis will reject the null hypothesis if the sample mean is significantly higher or lower than the hypothesized population mean. Thus a two-tailed test is appropriate when we are testing whether the population mean is different from some hypothesized value. A one-tailed test, on the other hand, would be used when we are testing whether the population mean is lower than or higher than some hypothesized value.

8-24 They should perform a lower-tailed test, with H_0: $\mu = 3124$, H_1: $\mu < 3124$.

8-26 $\sigma = 5.75$ $n = 25$ $\bar{x} = 42.95$

$H_0 : \mu = 44.95$ $H_1 : \mu < 44.95$ $\alpha = .02$

The lower limit of the acceptance region is $z_L = -2.05$, or

$$\bar{x}_L = \mu - z_{.02}\sigma/\sqrt{n} = 44.95 - 2.05(5.75)/\sqrt{25} = \$42.59$$

Since $z = \dfrac{\bar{x} - \mu}{\sigma/\sqrt{n}} = \dfrac{42.95 - 44.95}{5.75/\sqrt{25}} = -1.74 > -2.05$ (or $\bar{x} > 42.59$), we cannot reject H_0.

Atlas should not believe that the average retail price has decreased.

8-28 $\sigma = 18.4$ $n = 20$ $\bar{x} = 954$

$H_0 : \mu = 960$ $H_1 : \mu \neq 960$ $\alpha = .05$

The limits of the acceptance region are $z_{\text{CRIT}} = \pm 1.96$, or

$$\bar{x}_{\text{CRIT}} = \mu \pm z_{.025}\sigma/\sqrt{n} = 960 \pm 1.96(18.4)/\sqrt{20} = (951.94,\ 968.06) \text{ lumens}$$

Since $z = \dfrac{\bar{x} - \mu}{\sigma/\sqrt{n}} = \dfrac{954 - 960}{18.4/\sqrt{20}} = -1.46 > -1.96$ (or $\bar{x} > 951.94$), we do not reject H_0.

The average light output is not significantly different from the hypothesized value.

8-30 $\sigma = 52$ $n = 121$ $\bar{x} = 151$

$H_0 : \mu = 144$ $H_1 : \mu > 144$ $\alpha = .10$

The upper limit of the acceptance region is $z_U = 1.28$, or

$$\bar{x}_U = \mu + z_{.10}\sigma/\sqrt{n} = 144 + 1.28(52)/\sqrt{121} = \$150$$

Since $z = \dfrac{\bar{x} - \mu}{\sigma/\sqrt{n}} = \dfrac{151 - 144}{52/\sqrt{121}} = 1.48 > 1.28$, (or $\bar{x} > 150$), we should reject H_0. Joel's clients' average commission is significantly higher than the industry average.

8-32 $\sigma = 0.10$ $n = 15$ $\bar{x} = 0.33$

$H_0 : \mu = 0.57$ $H_1 : \mu < 0.57$ $\alpha = .01$

The lower limit of the acceptance region is $z_L = -2.33$, or

$$\bar{x}_L = \mu - z_{.01}\sigma/\sqrt{n} = 0.57 - 2.33(0.10)/\sqrt{15} = 0.51\%$$

Since $z = \dfrac{\bar{x} - \mu}{\sigma/\sqrt{n}} = \dfrac{0.33 - 0.57}{0.10/\sqrt{15}} = -9.30 < -2.33$ (or $\bar{x} < 0.51$), we should reject H_0. The rate of growth has decreased significantly, and we infer that this was because of the oil embargo and its consequences.

8-34 From exercise 8-31, we have $\sigma = 16$, $n = 64$, and $\bar{x}_{CRIT} = 31.28$.

a) $P(\bar{x} > 31.28 \mid \mu = 28) = P\left(z > \dfrac{31.28 - 28}{16/\sqrt{64}}\right) = P(z > 1.64) = .5 - .4495 = .0505$

b) $P(\bar{x} > 31.28 \mid \mu = 29) = P\left(z > \dfrac{31.28 - 29}{16/\sqrt{64}}\right) = P(z > 1.14) = .5 - .3729 = .1271$

c) $P(\bar{x} > 31.28 \mid \mu = 30) = P\left(z > \dfrac{31.28 - 30}{16/\sqrt{64}}\right) = P(z > .64) = .5 - .2389 = .2611$

8-36 From exercise 8-31, we have $\sigma = 16$, $n = 64$, $H_0 : \mu = 28$, $H_1 : \mu > 28$
At $\alpha = .02$, the upper limit of the acceptance region is

$$\mu + z_{.02}\sigma/\sqrt{n} = 28 + 2.05(16)/\sqrt{64} = 32.1 \text{ million dollars}$$

a) $P(\bar{x} > 32.1 \mid \mu = 28) = P\left(z > \dfrac{32.1 - 28}{16/\sqrt{64}}\right) = P(z > 2.05) = .5 - .4798 = .0202$

b) $P(\bar{x} > 32.1 \mid \mu = 29) = P\left(z > \dfrac{32.1 - 29}{16/\sqrt{64}}\right) = P(z > 1.55) = .5 - .4394 = .0606$

c) $P(\bar{x} > 32.1 \mid \mu = 30) = P\left(z > \dfrac{32.1 - 30}{16/\sqrt{64}}\right) = P(z > 1.05) = .5 - .3531 = .1469$

8-38 $n = 85$ $\bar{p} = .1412$
$H_0 : p = .19$ $H_1 : p < .19$ $\alpha = .04$
The lower limit of the acceptance region is $z_L = -1.75$, or

$$\bar{p}_L = p - z_{.04}\sqrt{\frac{pq}{n}} = .19 - 1.75\sqrt{\frac{.19(.81)}{85}} = .1155$$

Since $z = \dfrac{\bar{p} - p}{\sqrt{pq/n}} = \dfrac{.1412 - .19}{\sqrt{.19(.81)/85}} = -1.15 > -1.75$ (or $\bar{p} > .1155$), we don't reject H_0. Grant's western distribution is not significantly worse than its eastern distribution.

8-40 a) $n = 180$ $\bar{p} = 17/180 = .0944$
$H_0 : p = .151$ $H_1 : p < .151$ $\alpha = .05$

The lower limit of the acceptance region is $z_L = -1.64$, or

$$\bar{p}_L = p - z_{.05}\sqrt{\frac{pq}{n}} = .151 - 1.64\sqrt{\frac{.151(.849)}{180}} = .1072$$

Since $z = \dfrac{\bar{p} - p}{\sqrt{pq/n}} = \dfrac{.0944 - .151}{\sqrt{.151(.849)/180}} = -2.12 < -1.64$ (or $\bar{p} < .1072$), they should reject H_0. Spray users are significantly less susceptible to colds.

b) At $\alpha = .02$, the lower limit of the acceptance region is $z_L = -2.05$, or

$$\bar{p}_L = p - z_{.02}\sqrt{\frac{pq}{n}} = .151 - 2.05\sqrt{\frac{.151(.849)}{180}} = 0.0963$$

Again $z < -2.05$ ($\bar{p} < .0963$), so the same conclusion holds.

c) Not necessarily. Although the users of the spray seem to be significantly less susceptible to colds, we do not know that other relevant factors have been controlled for in the experiment, nor have we been told anything about potential side-effects of the spray which might counterbalance its effectiveness in reducing susceptibility to colds.

8-42 $n = 3000$ $\bar{p} = 950/3000 = .3167$
$H_0 : p = .35$ $H_1 : p < .35$ $\alpha = .05$
The lower limit of the acceptance region is $z_L = -1.645$, or

$$\bar{p}_L = p - z_{.05}\sqrt{\frac{pq}{n}} = .35 - 1.645\sqrt{\frac{.35(.65)}{3000}} = .3357$$

Since $z = \dfrac{\overline{p} - p}{\sqrt{pq/n}} = \dfrac{.3167 - .35}{\sqrt{.35(.65)/3000}} = -3.82 < -1.645$ (or $\overline{p} < .3357$), we should reject H_0.

The proportion of skeptical people is significantly less than it was last year.

8-44 $s = 8.4$ $n = 6$ $\overline{x} = 94.3$

$H_0 : \mu = 100$ $H_1 : \mu < 100$ $\alpha = .05$

The lower limit of the acceptance region is $t_L = -t_{5,\ .05} = -2.015$, or

$$\overline{x}_L = \mu - t_{5,\ .05} s/\sqrt{n} = 100 - 2.015(8.4)/\sqrt{6} = 93.09$$

Since $t = \dfrac{\overline{x} - \mu}{s/\sqrt{n}} = \dfrac{94.3 - 100}{8.4/\sqrt{6}} = -1.662 > -2.015$ (or $\overline{x} > 93.09$), we do not reject H_0.

8-46 $s = 49{,}000$ $n = 12$ $\overline{x} = 780{,}000$

$H_0 : \mu = 825{,}000$ $H_1 : \mu < 825{,}000$ $\alpha = .05$

The lower limit of the acceptance region is $t_L = -t_{11,\ .05} = -1.796$

$$\overline{x}_L = \mu - t_{11,\ .05} s/\sqrt{n} = 825{,}000 - 1.796(49{,}000)/\sqrt{12} = \$799{,}595$$

Since $t = \dfrac{\overline{x} - \mu}{s/\sqrt{n}} = \dfrac{780{,}000 - 825{,}000}{49{,}000/\sqrt{12}} = -3.181 < -1.796$ (or $\overline{x} < 799{,}595$), we reject H_0. The average appraised value of homes in the area is significantly less than \$825,000.

8-48 $s^2 = 16.2$ $(s = 4.025)$ $n = 95$ $\overline{x} = 7.2$

$H_0 : \mu = 8.1$ $H_1 : \mu < 8.1$ $\alpha = .01$

The lower limit of the acceptance region is $t_L = -t_{94,\ .01} = -2.33$, or

$$\overline{x}_L = \mu - t_{94,\ .01} s/\sqrt{n} = 8.1 - 2.33(4.025)/\sqrt{95} = 7.14 \text{ hours}$$

(with 94 *df*, we use the normal table)

Since $t = \dfrac{\overline{x} - \mu}{s/\sqrt{n}} = \dfrac{7.2 - 8.1}{4.025/\sqrt{95}} = -2.179 > -2.33$ (or $\overline{x} > 7.14$), we do not reject H_0. The new terminals are not significantly easier to learn to operate.

8-50 $s = 2.7$ $n = 18$ $\overline{x} = 12.4$

$H_0 : \mu = 10$ $H_1 : \mu \neq 10$ $\alpha = .01$

The limits of the acceptance region are $t_{crit} = \pm t_{17,\ .005} = \pm 2.898$, or

$$\overline{x}_{CRIT} = \mu \pm t_{17,\ .005} s/\sqrt{n} = 10 \pm 2.898(2.7)/\sqrt{18} = (8.16,\ 11.84) \text{ pounds}$$

Since $t = \dfrac{\overline{x} - \mu}{s/\sqrt{n}} = \dfrac{12.4 - 10}{2.7/\sqrt{18}} = 3.771 > 2.898$ (or $\overline{x} > 11.84$), we reject H_0. The claim does not appear to be valid.

8-52 a) Let p_{NY} = proportion of homeless people in New York City
p_{DC} = proportion of homeless people in Washington, D.C.
$H_0 : p_{NY} = p_{DC}$ $H_1 : p_{NY} \neq p_{DC}$

b) Let μ_A = mean sales after the promotion
μ_B = mean sales before the promotion
$H_0 : \mu_A = \mu_B$ $H_1 : \mu_A > \mu_B$

c) Let μ = average annual snowfall in the 1980's
H_0: $\mu = 8$ H_1: $\mu \neq 8$

d) Let μ = average mpg for the model
$H_0 : \mu = 34$ $H_1 : \mu < 34$

8-54 No, because each of the sample means is equally distant from the hypothesized mean and, therefore, equally likely to lead to rejection or acceptance of the null hypothesis.

8-56 From exercise 8-28, we have $\sigma = 2100$, $n = 25$, $H_0 : \mu = 14500$, $H_1 : \mu < 14500$
At $\alpha = .10$, the lower limit of the acceptance region is

$$\mu - z_{.10}\sigma/\sqrt{n} = 14500 - 1.28(2100)/\sqrt{25} = 13962.4 \text{ hours}$$

a) $P(\overline{x} < 13962.4 \mid \mu = 14000) = P\left(z < \dfrac{13962.4 - 14000}{2100/\sqrt{25}}\right)$

$= P(z < -0.09) = .5 - .0359 = .4641$

b) $P(\overline{x} < 13962.4 \mid \mu = 13500) = P\left(z < \dfrac{13962.4 - 13500}{2100/\sqrt{25}}\right)$

$= P(z < 1.10) = .5 + .3643 = .8643$

c) $P(\overline{x} < 13962.4 \mid \mu = 13000) = P\left(z < \dfrac{13962.4 - 13000}{2100/\sqrt{25}}\right)$

$= P(z < 2.29) = 5 + .4890 = .9890$

8-58 $n = 8000$ $\overline{p} = 18/8000 = .00225$

$H_0 : p = .003$ $H_1 : p < .003$ $\alpha = .10$
The lower limit of the acceptance region is $z_L = -1.28$, or

$$\overline{p}_L = p - z_{.10}\sqrt{\frac{pq}{n}} = .003 - 1.28\sqrt{\frac{.003(.997)}{8000}} = .00222$$

Since $z = \dfrac{\overline{p} - p}{\sqrt{pq/n}} = \dfrac{.00225 - .003}{\sqrt{.003(.997)/8000}} = -1.23 > -1.28$ (or $\overline{p} > .00222$), we do not reject H_0.

The new procedures have not significantly reduced the fraction of lost mail.

8-60 Let p = the proportion of closed-end equity funds selling at a discount.
$n = 15$ $\overline{p} = 6/15 = 0.4$
$H_0 : p = 0.5$ $H_1 : p < 0.5$ $\alpha = .01$
The lower limit of the acceptance region is $z_L = -2.33$, or

$$\overline{p}_L = p - z_{.01}\sqrt{\frac{pq}{n}} = 0.5 - 2.33\sqrt{\frac{.5(.5)}{15}} = 0.1992$$

Since $z = \dfrac{\overline{p} - p}{\sqrt{pq/n}} = \dfrac{.4 - .5}{\sqrt{.5(.5)/15}} = -0.77 > -2.33$ (or $\overline{p} > 0.1992$), we do not reject H_0.

The proportion of closed-end equity funds selling at a discount is not significantly less than the proportion selling at a premium.

8-62 a) $P(\text{accept} \mid H_0 \text{ true}) = .85 \Rightarrow \overline{x} \pm 1.44\sigma_{\overline{x}}$

b) $P(\text{accept} \mid H_0 \text{ true}) = .98 \Rightarrow \overline{x} \pm 2.33\sigma_{\overline{x}}$

8-64 $\sigma = 1.5$ $n = 200$ $\overline{x} = 31.7$
$H_0 : \mu = 32$ $H_1 : \mu < 32$ $\alpha = .02$
The lower limit of the acceptance region is $z_L = -2.05$, or

$$\overline{x}_L = \mu - z_{.02}\sigma/\sqrt{n} = 32 - 2.05(1.5)/\sqrt{200} = 31.7826 \text{ ounces}$$

Since $z = \dfrac{\overline{x} - \mu}{\sigma/\sqrt{n}} = \dfrac{31.7 - 32}{1.5/\sqrt{200}} = -2.83 < -2.05$ (or $\overline{x} < 31.7826$), we reject H_0. There is significant evidence that the bottles are being underfilled.

8-66 $N = 2400$ $n = 300$ $\bar{p} = 57/300 = .19$

$H_0: p = .15$ $H_1: p > .15$ $\alpha = .05$

The upper limit of the acceptance region is $z_U = 1.64$, or

$$\bar{p}_U = p + z_{.05}\sqrt{\frac{pq}{n}\left(\frac{N-n}{N-1}\right)} = .15 + 1.64\sqrt{\frac{.15(.85)}{300}\left(\frac{2100}{2399}\right)} = .1816$$

Since $z = \dfrac{\bar{p} - p}{\sqrt{\frac{pq}{n}\left(\frac{N-n}{N-1}\right)}} = \dfrac{.19 - .15}{\sqrt{\frac{.15(.85)}{300}\left(\frac{2100}{2399}\right)}} = 2.07 > 1.64$ (or $\bar{p} > .1816$), they should reject H_0

and open the store.

8-68 $n = 250$ $\bar{p} = 194/250 = .7760$

$H_0: p = .72$ $H_1: p > .72$ $\alpha = .02$

The upper limit of the acceptance region is $z_U = 2.05$, or

$$\bar{p}_U = p + z_{.02}\sqrt{\frac{pq}{n}} = .72 + 2.05\sqrt{\frac{.72(.28)}{250}} = .7782$$

Since $z = \dfrac{\bar{p} - p}{\sqrt{pq/n}} = \dfrac{.7760 - .72}{\sqrt{.72(.28)/250}} = 1.97 < 2.05$ (or $\bar{p} = .7760 < .7782$), we do not reject H_0.

The survey does not support the editor's belief.

8-70 $s = 19.48$ $n = 18$ $\bar{x} = 87.61$

$H_0: \mu = 77.38$ $H_1: \mu > 77.38$ $\alpha = .025$

The upper limit of the acceptance region is $t_U = t_{17,\ .025} = 2.110$, or

$$\bar{x}_U = \mu + t_{17,.025}s/\sqrt{n} = 77.38 + 2.110(19.48)/\sqrt{18} = \$87.07$$

Since $t = \dfrac{\bar{x} - \mu}{s/\sqrt{n}} = \dfrac{87.61 - 77.38}{19.48/\sqrt{18}} = 2.28 > 2.110$ (or $\bar{x} > 87.07$), we reject H_0 and conclude that

Drive-a-Lemon's average total charge is significantly higher than the average total charge at the major national chains. However, this need not indicate that Drive-a-Lemon's rates are not lower than the rates of the major chains. For example, suppose Drive-a-Lemon has offices only in New York and Chicago. The \$77.38 average total charge established by the survey includes charges incurred in other areas which are significantly less expensive than New York and Chicago, and these charges could account for the low average. To validate its claim, Drive-a-Lemon should look at average total charges for the national chains on rentals made in Drive-a-Lemon's service area.

8-72 $n = 20$ $\bar{p} = 12/20 = .6$

$H_0: p = .5$ $H_1: p \neq .5$ $\alpha = .10$

The limits of the acceptance region are $z_{CRIT} = \pm 1.64$, or

$$\bar{p}_{CRIT} = p \pm z_{.05}\sqrt{\frac{pq}{n}} = .5 \pm 1.64\sqrt{\frac{.5(.5)}{20}} = (0.3166,\ 0.6834)$$

Since $z = \dfrac{\bar{p} - p}{\sqrt{pq/n}} = \dfrac{.6 - .5}{\sqrt{.5(.5)/20}} = 0.8944 < 1.64$ (or $\bar{p} = .6 < 0.6834$), we fail to reject H_0.

The proportion of CEOs' families with more than two children is not significantly different from 0.5.

8-74 From exercise 8-26, we have $\sigma = 5.75$, $n = 25$, and $\bar{x}_{CRIT} = 42.59$.

a) $P(\bar{x} \le 42.59 \mid \mu = 41.95) = P\left(z \le \frac{42.59 - 41.95}{5.75/\sqrt{25}}\right)$

$= P(z \le .56) = .5 + .2123 = .7123$

b) $P(\bar{x} \le 42.59 \mid \mu = 42.95) = P\left(z \le \frac{42.59 - 42.95}{5.75/\sqrt{25}}\right)$

$= P(z \le -.31) = .5 - .1217 = .3783$

c) $P(\bar{x} \le 42.59 \mid \mu = 43.95) = P\left(z \le \frac{42.59 - 43.95}{5.75/\sqrt{25}}\right)$

$= P(z \le -1.18) = 5 - .3810 = .1190$

8-76 From exercise 8-28, we have $\sigma = 2100$, $n = 25$, and $\bar{x}_{CRIT} = 13521.4$.

a) $P(\bar{x} \le 13521.4 \mid \mu = 14000) = P\left(z \le \frac{13521.4 - 14000}{2100/\sqrt{25}}\right)$

$= P(z \le -1.14) = .5 - .3729 = .1271$

b) $P(\bar{x} \le 13521.4 \mid \mu = 13500) = P\left(z \le \frac{13521.4 - 13500}{2100/\sqrt{25}}\right)$

$= P(z \le .05) = .5 + .0199 = .5199$

c) $P(\bar{x} \le 13521.4 \mid \mu = 13000) = P\left(z \le \frac{13521.4 - 13000}{2100/\sqrt{25}}\right)$

$= P(z \le 1.24) = 5 + .3925 = .8925$

8-78 From exercise 8-26, we have $\sigma = 5.75$, $n = 25$, $H_0: \mu = 44.95$, $H_1: \mu < 44.95$
At $\alpha = .05$, the lower limit of the acceptance region is

$$\mu - z_{.05}\sigma/\sqrt{n} = 44.95 - 1.64(5.75)/\sqrt{25} = \$43.06$$

a) $P(\bar{x} \le 43.06 \mid \mu = 41.95) = P\left(z \le \frac{43.06 - 41.95}{5.75/\sqrt{25}}\right)$

$= P(z \le .97) = .5 + .3340 = .8340$

b) $P(\bar{x} \le 43.06 \mid \mu = 42.95) = P\left(z \le \frac{43.06 - 42.95}{5.75/\sqrt{25}}\right)$

$= P(z \le .10) = .5 + .0398 = .5398$

c) $P(\bar{x} \le 43.06 \mid \mu = 43.95) = P\left(z \le \frac{43.06 - 43.95}{5.75/\sqrt{25}}\right)$

$= P(z \le -.77) = 5 - .2794 = .2206$

8-80 $N = 5000$ $n = 500$ $\bar{p} = .43$
$H_0: p = .48$ $H_1: p < .48$ $\alpha = .01$

The lower limit of the acceptance region is $z_L = -2.33$, or

$$\bar{p}_L = p - z_{.01}\sqrt{\frac{pq}{n}\left(\frac{N-n}{N-1}\right)} = .48 - 2.33\sqrt{\frac{.48(.52)}{500}\left(\frac{4500}{4999}\right)} = .4306$$

Since $z = \dfrac{\overline{p} - p}{\sqrt{\dfrac{pq}{n}\left(\dfrac{N-n}{N-1}\right)}} = \dfrac{.43 - .48}{\sqrt{\dfrac{.48(.52)}{500}\left(\dfrac{4500}{4999}\right)}} = -2.36 < -2.33$ (or $\overline{p} < .4306$), we reject H_0. The company has fallen significantly below its target of a 48% market share.

CHAPTER 9

TESTING HYPOTHESES: TWO - SAMPLE TESTS

9-2 Presumably they wish to determine if $\mu_S < \mu_A$, but, of course, even if this is true, they should also be concerned about the relative costs of the two lines.

$s_S = 32$ $n_S = 150$ $\bar{x}_S = 198$ $s_A = 29$ $n_A = 200$ $\bar{x}_A = 206$

$H_0: \mu_S = \mu_A$ $H_1: \mu_S < \mu_A$ $\alpha = .02$

$$\hat{\sigma}_{\bar{x}_S - \bar{x}_A} = \sqrt{\frac{s_S^2}{n_S} + \frac{s_A^2}{n_A}} = \sqrt{\frac{(32)^2}{150} + \frac{(29)^2}{200}} = 3.3214 \text{ chips per hour}$$

The lower limit of the acceptance region is $z_L = -2.05$, or

$$(\bar{x}_S - \bar{x}_A)_L = 0 - z_{.02}\hat{\sigma}_{\bar{x}_S - \bar{x}_A} = -2.05(3.3214) = -6.8089 \text{ chips per hour}$$

Since $z = \dfrac{(\bar{x}_S - \bar{x}_A) - (\bar{\mu}_S - \bar{\mu}_A)}{\hat{\sigma}_{\bar{x}_S - \bar{x}_A}} = \dfrac{(198 - 206) - 0}{3.3214} = -2.41 < -2.05$

(or $\bar{x}_S - \bar{x}_A = -8 < -6.8089$), we reject H_0. The output from the automatic line is significantly greater than that from the semiautomatic line.

9-4 Sample 1 (1992): $s_1 = 0.84$ $n_1 = 38$ $\bar{x}_1 = 4.36$

Sample 2 (1993): $s_2 = 0.51$ $n_2 = 32$ $\bar{x}_2 = 3.23$

$H_0: \mu_1 = \mu_2$ $H_1: \mu_1 > \mu_2$ $\alpha = .05$

$$\hat{\sigma}_{\bar{x}_1 - \bar{x}_2} = \sqrt{\frac{s_1^2}{n_1} + \frac{s_2^2}{n_2}} = \sqrt{\frac{(0.84)^2}{38} + \frac{(0.51)^2}{32}} = 0.1634\%$$

The upper limit of the acceptance region is $z_U = 1.64$, or

$$(\bar{x}_1 - \bar{x}_2)_{CRIT} = 0 + z_{.05}\hat{\sigma}_{\bar{x}_1 - \bar{x}_2} = 1.64(.1634) = 0.2680\%$$

Since $z = \dfrac{(\bar{x}_1 - \bar{x}_2) - (\bar{\mu}_1 - \bar{\mu}_2)}{\hat{\sigma}_{\bar{x}_1 - \bar{x}_2}} = \dfrac{(4.36 - 3.23) - 0}{0.1634} = 6.92 > 1.64$ (or $\bar{x}_1 - \bar{x}_2 = 1.13\% > 0.2680\%$), we reject H_0 and conclude that money market rates declined significantly in 1992.

9-6 Sample 1 (male): $s_1 = 1.84$ $n_1 = 38$ $\bar{x}_1 = 11.38$

Sample 2 (female): $s_2 = 1.31$ $n_2 = 45$ $\bar{x}_2 = 8.42$

$H_0: \mu_1 - \mu_2 = 2$ $H_1: \mu_1 - \mu_2 > 2$ $\alpha = .01$

$$\hat{\sigma}_{\bar{x}_1 - \bar{x}_2} = \sqrt{\frac{s_1^2}{n_1} + \frac{s_2^2}{n_2}} = \sqrt{\frac{(1.84)^2}{38} + \frac{(1.31)^2}{45}} = \$0.3567$$

The upper limit of the acceptance region is $z_U = 2.33$, or

$$(\bar{x}_1 - \bar{x}_2)_U = (\mu_1 - \mu_2)_{H_0} + z_{.01}\hat{\sigma}_{\bar{x}_1 - \bar{x}_2} = 2 + 2.33(.3567) = \$2.8311$$

Since $z = \dfrac{(\bar{x}_1 - \bar{x}_2) - (\bar{\mu}_1 - \bar{\mu}_2)}{\hat{\sigma}_{\bar{x}_1 - \bar{x}_2}} = \dfrac{(11.38 - 8.42) - 2}{0.3567} = 2.69 > 2.33$ (or $\bar{x}_1 - \bar{x}_2 = 2.96 > 2.831$), we reject H_0 and conclude that male operators do earn over \$2.00 more per hour than female operators.

9-8 $s_O = 32.63 \quad n_O = 16 \quad \bar{x}_O = 688 \quad s_N = 24.84 \quad n_N = 11 \quad \bar{x}_N = 706$
$H_0 : \mu_O = \mu_N \quad H_1 : \mu_O < \mu_N \quad \alpha = .05$

$$s_p = \sqrt{\frac{(n_O - 1)s_O^2 + (n_N - 1)s_N^2}{n_O + n_N - 2}} = \sqrt{\frac{15(32.63)^2 + 10(24.84)^2}{25}} = \$29.7597$$

The lower limit of the acceptance region is $t_L = -t_{25,\ .05} = -1.708$, or

$$(\bar{x}_A - \bar{x}_N)_L = 0 - t_{25,\ .05}s_p\sqrt{\frac{1}{n_O} + \frac{1}{n_N}} = -1.708(29.7597)\sqrt{\frac{1}{16} + \frac{1}{11}} = -\$19.91$$

Since $t = \dfrac{(\bar{x}_O - \bar{x}_B) - (\bar{\mu}_O - \bar{\mu}_N)}{s_p\sqrt{\frac{1}{n_O} + \frac{1}{n_N}}} = \dfrac{(688 - 706) - 0}{29.7597\sqrt{\frac{1}{16} + \frac{1}{11}}} = -1.544 > -1.708$ (or $\bar{x}_O - \bar{x}_B = -18 > -19.91$), we do not reject H_0. Average daily sales have not increased significantly.

9-10 $s_1 = 370 \quad n_1 = 9 \quad \bar{x}_1 = 2990 \quad s_2 = 805 \quad n_2 = 6 \quad \bar{x}_2 = 3065$
$H_0 : \mu_1 = \mu_2 \quad H_1 : \mu_1 < \mu_2 \quad \alpha = .05$

$$s_p = \sqrt{\frac{(n_1 - 1)s_1^2 + (n_2 - 1)s_2^2}{n_1 + n_2 - 2}} = \sqrt{\frac{8(370)^2 + 5(805)^2}{13}} = \$577.48$$

The lower limit of the acceptance region is $t_L = -t_{13,\ .05} = -1.771$, or

$$(\bar{x}_1 - \bar{x}_2)_L = 0 - t_{13,\ .05}s_p\sqrt{\frac{1}{n_1} + \frac{1}{n_2}} = -1.771(577.48)\sqrt{\frac{1}{9} + \frac{1}{6}} = -\$539.02$$

Since $t = \dfrac{(\bar{x}_1 - \bar{x}_2) - (\bar{\mu}_1 - \bar{\mu}_2)}{s_p\sqrt{\frac{1}{n_1} + \frac{1}{n_2}}} = \dfrac{(2990 - 3065) - 0}{577.48\sqrt{\frac{1}{9} + \frac{1}{6}}} = -0.246 > -1.771$ (or $\bar{x}_1 - \bar{x}_2 = -75 > -539.02$), we do not reject H_0. The pear-shaped stones are not significantly more expensive than the marquise stones.

9-12 Sample 1 (mail): $s_m = 378 \quad n_m = 17 \quad \bar{x}_m = 563$
Sample 2 (electronic): $s_e = 619 \quad n_e = 13 \quad \bar{x}_e = 958$
$H_0 : \mu_m = \mu_e \quad H_1 : \mu_m < \mu_e \quad \alpha = .01$

$$s_p = \sqrt{\frac{(n_m - 1)s_m^2 + (n_e - 1)s_e^2}{n_m + n_e - 2}} = \sqrt{\frac{16(378)^2 + 12(619)^2}{28}} = \$495.84$$

The lower limit of the acceptance region is $t_L = -t_{28,\ .01} = -2.467$, or

$$(\bar{x}_m - \bar{x}_e)_L = (\mu_m - \mu_e)_{H_0} - t_{28,\ .01}s_p\sqrt{\frac{1}{n_m} + \frac{1}{n_e}} = 0 - (-2.467)(495.84)\sqrt{\frac{1}{17} + \frac{1}{13}} = -450.69$$

Since $t = \dfrac{(\bar{x}_m - \bar{x}_e) - (\bar{\mu}_m - \bar{\mu}_e)}{S_p\sqrt{\frac{1}{n_m} + \frac{1}{n_e}}} = \dfrac{(563 - 958) - 0}{495.84\sqrt{\frac{1}{17} + \frac{1}{13}}} = -2.162 > -2.162$ (or $\bar{x}_m - \bar{x}_e = -395 > -450.69$), we do not reject H_0. Refunds filed by mail were not significantly smaller than those filed electronically.

9-14

Firm	1	2	3	4	5	6	7	8	9
1991	1.38	1.26	3.64	3.50	2.47	3.21	1.05	1.98	2.72
1992	2.48	1.50	4.59	3.06	2.11	2.80	1.59	0.92	0.47
Change	1.10	0.24	0.95	−0.44	−0.36	−0.41	0.54	−1.06	−2.25

a) $\bar{x} = \frac{\sum x}{n} = \frac{-1.69}{9} = -0.1878$

b) $s^2 = \frac{1}{n-1}\left(\sum x^2 - n\bar{x}^2\right) = \frac{1}{8}\left(9.1391 - 9(-0.1878)^2\right) = 1.1027$

$s = \sqrt{s^2} = 1.0501$

$\hat{\sigma}_{\bar{x}} = s/\sqrt{n} = 1.0501/\sqrt{9} = 0.3500$

c) $H_0 : \mu_{1991} = \mu_{1992}$ $\qquad H_1 : \mu_{1991} \neq \mu_{1992}$ $\qquad \alpha = .02$

The limits of the acceptance region are $t_{CRIT} = \pm t_{8,\ .01} = \pm 2.896$, or

$$\bar{x}_{CRIT} = 0 \pm t_{8,.01}\hat{\sigma}_{\bar{x}} = \pm\ 2.896(0.3500) = \pm\ 1.0136$$

Since $t = \frac{\bar{x}-\mu}{\hat{\sigma}_{\bar{x}}} = \frac{-0.1878 - 0}{0.3500} = -0.537 > -2.896$ (or $\bar{x} = -0.1878$), we do not reject H_0.

Average earnings per share did not change significantly from 1991 to 1992.

9-16

Pair	1	2	3	4	5	6	7	8	9
Regular	5.7	6.1	5.9	6.2	6.4	5.1	5.9	6.0	5.5
Additive	6.0	6.2	5.8	6.6	6.7	5.3	5.7	6.1	5.9
Difference	0.3	0.1	-0.1	0.4	0.3	0.2	-0.2	0.1	0.4

$\bar{x} = \frac{\sum x}{n} = \frac{1.5}{9} = 0.1667$ mpg

$s^2 = \frac{1}{n-1}\left(\sum x^2 - n\bar{x}^2\right) = \frac{1}{8}\left(0.61 - 9(0.1667)^2\right) = 0.0450,\quad s = \sqrt{s^2} = 0.2121$ mpg

$\hat{\sigma}_{\bar{x}} = s/\sqrt{n} = 0.2121/\sqrt{9} = 0.0707$ mpg

$H_0 : \mu_A = \mu_R$ $\qquad H_1 : \mu_A > \mu_R$ $\qquad \alpha = .01$

The upper limit of the acceptance region is $t_U = t_{8,\ .01} = 2.896$, or

$$\bar{x}_U = 0 + t_{8,.01}\hat{\sigma}_{\bar{x}} = 2.896(0.0707) = 0.2047 \text{ mpg}$$

Since $t = \frac{\bar{x}-\mu}{\hat{\sigma}_{\bar{x}}} = \frac{0.1667 - 0}{0.0707} = 2.358 < 2.896$ (or $\bar{x} = 0.1667$), we do not reject H_0. The additive does not yield significantly better fuel efficiency.

9-18

Employee	1	2	3	4	5	6
Production without music	219	205	226	198	209	216
Production with music	235	186	240	203	221	205
Change in production	16	−19	14	5	12	−11

$\bar{x} = \frac{\sum x}{n} = \frac{17}{6} = 2.8333$

$s^2 = \frac{1}{n-1}\left(\sum x^2 - n\bar{x}^2\right) = \frac{1}{5}\left(1103 - 6(2.8333)^2\right) = 210.9667,\ s = \sqrt{s^2} = 14.5247$

$\hat{\sigma}_{\bar{x}} = s/\sqrt{n} = 14.5247/\sqrt{6} = 5.9297$

$H_0 : \mu_{AFTER} = \mu_{BEFORE}$ $\qquad H_1 : \mu_{AFTER} \neq \mu_{BEFORE}$ $\qquad \alpha = .02$

The limits of the acceptance region are $t_{CRIT} = \pm t_{5,\ .01} = \pm 3.365$, or

$$\bar{x}_{CRIT} = 0 \pm t_{5,\ .01}\hat{\sigma}_{\bar{x}} = \pm\ 3.365(5.9297) = \pm\ 19.95$$

Since $t = \dfrac{\bar{x} - \mu}{\hat{\sigma}_{\bar{x}}} = \dfrac{2.8333 - 0}{5.9297} = 0.478 < 3.365$ (or $\bar{x} = 2.8333$), we do not reject H_0. The music does not have a significant effect on productivity.

9-20 $n_1 = 40$ $\quad \bar{p}_1 = .275$ $\quad n_2 = 60$ $\quad \bar{p}_2 = .40$

$H_0 : p_1 = p_2$ $\quad H_1 : p_1 < p_2$ $\quad \alpha = .10$

$$\hat{p} = \frac{n_1\bar{p}_1 + n_2\bar{p}_2}{n_1 + n_2} = \frac{40(.275) + 60(.40)}{40 + 60} = 0.35$$

$$\hat{\sigma}_{\bar{p}_1 - \bar{p}_2} = \sqrt{\hat{p}\hat{q}\left(\frac{1}{n_1} + \frac{1}{n_2}\right)} = \sqrt{.35(.65)\left(\frac{1}{40} + \frac{1}{60}\right)} = 0.0974$$

The lower limit of the acceptance region is $z_L = -z_{.10} = -1.28$, or

$$(\bar{p}_1 - \bar{p}_2)_L = 0 - z_{.10}\hat{\sigma}_{\bar{p}_1 - \bar{p}_2} = -1.28(.0974) = -0.1247$$

Since $z = \dfrac{\bar{p}_1 - \bar{p}_2}{\hat{\sigma}_{\bar{p}_1 - \bar{p}_2}} = \dfrac{0.275 - 0.4}{0.0974} = -1.283 < -1.28$ (or $\bar{p}_1 - \bar{p}_2 = -.125 < -.1247$), we reject H_0. The proportion of NYSE stocks that advanced on Friday was significantly smaller than those that advanced on Thursday (but just barely significant).

9-22 $n_1 = 200$ $\quad \bar{p}_1 = .68$ $\quad n_2 = 250$ $\quad \bar{p}_2 = .76$

$H_0 : p_1 = p_2$ $\quad H_1 = p_1 < p_2$ $\quad \alpha = .02$

$$\hat{p} = \frac{n_1\bar{p}_1 + n_2\bar{p}_2}{n_1 + n_2} = \frac{200(.68) + 250(.76)}{200 + 250} = 0.7244$$

$$\hat{\sigma}_{\bar{p}_1 - \bar{p}_2} = \sqrt{\hat{p}\hat{q}\left(\frac{1}{n_1} + \frac{1}{n_2}\right)} = \sqrt{.7244(.2756)\left(\frac{1}{200} + \frac{1}{250}\right)} = 0.0424$$

The lower limit of the acceptance region is $z_L = -z_{.02} = -2.05$, or

$$(\bar{p}_1 - \bar{p}_2)_L = 0 - z_{.02}\hat{\sigma}_{\bar{p}_1 - \bar{p}_2} = -2.05(0.0424) = -0.0869$$

Since $z = \dfrac{\bar{p}_1 - \bar{p}_2}{\hat{\sigma}_{\bar{p}_1 - \bar{p}_2}} = \dfrac{0.68 - 0.76}{0.0424} = -1.89 > -2.05$ (or $\bar{p}_1 - \bar{p}_2 = -0.08 > -0.0869$), we do not reject H_0. The less expensive system will be installed.

9-24 $n_F = 150$ $\quad \bar{p}_F = .46$ $\quad n_S = 175$ $\quad \bar{p}_S = .40$

$H_0 : p_F = p_S$ $\quad H_F : p_F \neq p_S$ $\quad \alpha = .10$

$$\hat{p} = \frac{n_F\bar{p}_F + n_S\bar{p}_S}{n_F + n_S} = \frac{150(.46) + 175(.40)}{150 + 175} = .4277$$

$$\hat{\sigma}_{\bar{p}_F - \bar{p}_S} = \sqrt{\hat{p}\hat{q}\left(\frac{1}{n_F} + \frac{1}{n_S}\right)} = \sqrt{.4277(.5723)\left(\frac{1}{150} + \frac{1}{175}\right)} = .0551$$

The limits of the acceptance region are $z_{CRIT} = \pm z_{.05} = \pm 1.64$, or

$$(\bar{p}_F - \bar{p}_S)_{CRIT} = 0 \pm z_{.05}\hat{\sigma}_{\bar{p}_F - \bar{p}_S} = \pm\ 1.64(.0551) = \pm\ .0904$$

Since $z = \dfrac{\bar{p}_F - \bar{p}_S}{\hat{\sigma}_{\bar{p}_F - \bar{p}_S}} = \dfrac{0.46 - 0.40}{0.0551} = 1.09 < 1.64$ (or $\bar{p}_F - \bar{p}_S = 0.06 < 0.0904$), we do not reject H_0. The proportions of freshmen and sophomores do not differ significantly.

9-26 $\sigma = 7600$ $\quad n = 64$ $\quad \bar{x} = 38500$

$H_0 : \mu = 40000$ $\quad H_1 : \mu < 40000$

$$P(\bar{x} \leq 38500 \mid H_0) = P\left(z \leq \frac{38500 - 40000}{7600/\sqrt{64}}\right) = P(z \leq -1.58) = .5 - .4429 = .0571$$

9-28 $\sigma = .06$ $\quad n = 25$ $\quad \bar{x} = 4.97$

$H_0 : \mu = 5.00$ $\quad H_1 : \mu \neq 5.00$

$$P(\bar{x} \leq 4.97 \text{ or } \bar{x} \geq 5.03 \mid H_0) = 2P\left(z \geq \frac{5.03 - 5.00}{.06/\sqrt{25}}\right) = 2P(z \geq 2.5) = 2(.5 - .4938) = .0124$$

Hence, for any significance level greater than .0124 we should reject H_0 (and recalibrate the machine), but if $\alpha < .0124$, we will accept H_0 (and leave the machine as it is currently set).

9-30 From exercise 9-2, we have $s_1 = 32$, $n_1 = 150$, $\bar{x}_1 = 198$, and $s_2 = 29$, $n_2 = 200$, $\bar{x}_2 = 206$

$$\hat{\sigma}_{\bar{x}_1 - \bar{x}_2} = \sqrt{\frac{s_1^2}{n_1} + \frac{s_2^2}{n_2}} = \sqrt{\frac{(32)^2}{150} + \frac{(29)^2}{200}} = 3.3214$$

$$z = \frac{(\bar{x}_1 - \bar{x}_2) - (\mu_1 - \mu_2)_{H_0}}{\hat{\sigma}_{\bar{x}_1 - \bar{x}_2}} = \frac{(198 - 206) - 0}{3.3214} = -2.409$$

$$P(z \leq -2.409) = .5 - .4920 = .0080$$

9-32 From exercise 9-8, we have $t = -1.544$ (which is greater than -1.708), so the probability value is greater than .05 in a lowered-tailed test with 25 degrees of freedom.

9-34 From exercise 9-14, we have $s = 1.0501$, $n = 9$, $\bar{x} = -0.1878$

$$\hat{\sigma}_{\bar{x}} = \frac{s}{\sqrt{n}} = \frac{1.0501}{\sqrt{9}} = 0.35$$

$$t = \frac{\bar{x} - \mu}{\hat{\sigma}_{\bar{x}}} = \frac{-0.1878 - 0}{0.35} = -0.5366$$

Since $t = -0.537$ (which is greater than -1.860 in a two-tailed test with 8 degrees of freedom), the probaility value is greater than 0.10.

9-36 From exercise 9-22, we have $n_1 = 200$, $\bar{p}_1 = .68$, $n_2 = 250$, $\bar{p}_2 = .76$, $\hat{\sigma}_{\bar{p}_1 - \bar{p}_2} = 0\ .0424$

$$z = \frac{\bar{p}_1 - \bar{p}_2}{\hat{\sigma}_{\bar{p}_1 - \bar{p}_2}} = \frac{.68 - .76}{.0424} = -1.887$$

Since $z = -1.887$, in a lower tailed test the probablity value is $P(z \leq -1.89) = .5 - .4706 = 0.0294$

9-38 $s_S = 8$ $\quad n_S = 40$ $\quad \bar{x}_S = 42$ $\quad s_F = 7$ $\quad n_F = 40$ $\quad \bar{x}_F = 45$

$H_0 : \mu_S = \mu_F$ $\quad H_1 : \mu_S \neq \mu_F$ $\quad \alpha = .02$

$$\hat{\sigma}_{\bar{x}_S - \bar{x}_F} = \sqrt{\frac{s_S^2}{n_S} + \frac{s_F^2}{n_F}} = \sqrt{\frac{8^2}{40} + \frac{7^2}{40}} = 1.6808 \text{ pens per store per month}$$

The limits of the acceptance region are $z_{CRIT} = \pm z_{.01} = \pm 2.33$, or

$$(\bar{x}_S - \bar{x}_F)_{CRIT} = 0 \pm z_{.01}\hat{\sigma}_{\bar{x}_S - \bar{x}_F} = \pm\ 2.33(1.6808) = \pm\ 3.9163 \text{ pens per store per month}$$

Since $z = \dfrac{(\bar{x}_S - \bar{x}_F) - (\bar{\mu}_S - \bar{\mu}_F)}{\hat{\sigma}_{\bar{x}_S - \bar{x}_F}} = \dfrac{(42 - 45) - 0}{1.6808} = -1.78 > -2.33$ (or $\bar{x}_S - \bar{x}_F = -3 > -3.9163$), we do not reject H_0. The two displays did not result in significantly different sales levels.

9-40 a) Before sample: $n = 11$ $\qquad \sum x = 195$ $\qquad \sum x^2 = 4195$

$$\bar{x}_B = \frac{\sum x}{n} = \frac{195}{11} = 17.7273 \text{ ounces}$$

$$s_B^2 = \frac{1}{n-1}\left(\sum x^2 - n\bar{x}^2\right) = \frac{1}{10}\left(4195 - 11(17.7273)^2\right) = 73.8171$$

After sample: $n = 11$ $\qquad \sum x = 241$ $\qquad \sum x^2 = 5987$

$$\bar{x}_A = \frac{241}{11} = 21.9091 \text{ ounces}$$

$$s_A^2 = \frac{1}{10}\left(5987 - 11(21.9091)^2\right) = 70.6905$$

$$s_p = \sqrt{\frac{(n_B - 1)s_B^2 + (n_A - 1)s_A^2}{n_B + n_A - 2}} = \sqrt{\frac{10(73.8171) + 10(70.6905)}{20}} = 8.5002 \text{ oz}$$

$$\hat{\sigma}_{\bar{x}_A - \bar{x}_B} = s_p\sqrt{\frac{1}{n_B} + \frac{1}{n_A}} = 8.5002\sqrt{\frac{1}{11} + \frac{1}{11}} = 3.6245 \text{ ounces}$$

$H_0 : \mu_A = \mu_B \qquad H_1 : \mu_A > \mu_B \qquad \alpha = .05$

The upper limit of the acceptance region is $t_U = t_{20,\ .05} = 1.725$, or

$$(\bar{x}_A - \bar{x}_B)_U = 0 + t_{20,.05}\hat{\sigma}_{\bar{x}_A - \bar{x}_B} = 1.725(3.6245) = 6.2523 \text{ ounces}$$

Since $t = \dfrac{(\bar{x}_A - \bar{x}_B) - (\bar{\mu}_A - \bar{\mu}_B)}{\hat{\sigma}_{\bar{x}_A - \bar{x}_B}} = \dfrac{(21.9091 - 17.7273) - 0}{3.6245} = 1.154 < 1.725$ (or $\bar{x}_A - \bar{x}_B = 4.1818 < 6.2523$), we do not reject H_0. The demand has not increased significantly.

b) A better sampling procedure would be to re-interview the same 11 customers who were surveyed before the campaign. Since this would control for other factors, he would expect to see a smaller value of $\hat{\sigma}_{\bar{x}_A - \bar{x}_B}$, so any observed difference would be more likely to be significant.

9-42 $n_B = 150 \qquad \bar{p}_B = .44 \qquad n_A = 200 \qquad \bar{p}_A = .52$

$H_0 : \bar{p}_A = \bar{p}_B \qquad H_1 : \bar{p}_A > \bar{p}_B \qquad \alpha = .04$

$$\hat{p} = \frac{n_A\bar{p}_A + n_B\bar{p}_B}{n_A + n_B} = \frac{200(.52) + 150(.44)}{200 + 150} = .4857$$

$$\hat{\sigma}_{\bar{p}_A - \bar{p}_B} = \sqrt{\hat{p}\hat{q}\left(\frac{1}{n_A} + \frac{1}{n_B}\right)} = \sqrt{.4857(.5143)\left(\frac{1}{200} + \frac{1}{150}\right)} = .0540$$

The upper limit of the acceptance region is $z_U = z_{.04} = 1.75$, or

$$(\bar{p}_A - \bar{p}_B)_U = 0 + z_{.04}\hat{\sigma}_{\bar{p}_A - \bar{p}_B} = 1.75(.0540) = .0945$$

Since $z = \dfrac{\bar{p}_A - \bar{p}_B}{\hat{\sigma}_{\bar{p}_A - \bar{p}_B}} = \dfrac{.52 - .44}{.0540} = 1.48 < 1.75$ (or $\bar{p}_A - \bar{p}_B = 0.08 < 0.0945$), we do not reject H_0. The campaign was not significantly effective.

9-44

Buffered Aspirin	16.5	25.5	23.0	14.5	28.0	10.0	21.5	18.5	15.5
Competition	12.0	20.5	25.0	16.5	24.0	11.5	17.0	15.0	13.0
Difference	−4.5	−5.0	2.0	2.0	−4.0	1.5	−4.5	−3.5	−2.5

$$\bar{x} = \frac{\sum x}{n} = \frac{-18.5}{9} = -2.0556$$

$$s^2 = \frac{1}{n-1}\left(\sum x^2 - n\bar{x}^2\right) = \frac{1}{8}\left(110.25 - 9(-2.0556)^2\right) = 9.0276$$

$s = \sqrt{s^2} = 3.0046$ minutes
$\hat{\sigma}_{\overline{x}} = s/\sqrt{n} = 3.0046/\sqrt{9} = 1.0015$ minutes
$H_0 : \mu_C = \mu_B$ $\qquad$ $H_1 : \mu_C > \mu_B$ $\qquad$ $\alpha = .10$
The limits of the acceptance region are $t_{CRIT} = \pm t_{8,\ .05} = \pm 1.860$, or

$$\overline{x}_{CRIT} = 0 \pm t_{8,\ .05}\hat{\sigma}_{\overline{x}} = \pm\ 1.860(1.0015) = \pm\ 1.8628 \text{ minutes}$$

Since $t = \dfrac{\overline{x} - \mu}{\hat{\sigma}_{\overline{x}}} = \dfrac{-2.0556 - 0}{1.0015} = 2.053 > 1.860$ (or $\overline{x} = -2.0556 < -1.860$), we do reject H_0.

There is a significant difference in the times the two medications take to reach the bloodstream.

9-46

Old Formula	5	2	5	4	3	6	2	4	2	6	5	7	1	3
New Formula	3	1	5	1	1	4	4	2	5	2	3	3	1	2
Difference	−2	−1	0	−3	−2	−2	2	−2	3	−4	−2	−4	0	−1

$\overline{x} = \dfrac{\sum x}{n} = \dfrac{-18}{14} = -1.2857$ days

$s^2 = \dfrac{1}{n-1}\left(\sum x^2 - n\overline{x}^2\right) = \dfrac{1}{13}\left(76 - 14(-1.2857)^2\right) = 4.0660,\ s = \sqrt{s^2} = 2.0164$

$\hat{\sigma}_{\overline{x}} = s/\sqrt{n} = 2.0164/\sqrt{14} = 0.5389$

$H_0 : \mu_{NEW} = \mu_{OLD}$ $\qquad$ $H_1 : \mu_{NEW} < \mu_{OLD}$ $\qquad$ $\alpha = .01$
The lower limit of the acceptance region is $t_L = -t_{13,\ .01} = -2.650$, or

$$\overline{x}_L = 0 - t_{13,\ .01}\hat{\sigma}_{\overline{x}} = -2.650(0.5389) = -1.4281$$

Since $t = \dfrac{\overline{x} - \mu}{\hat{\sigma}_{\overline{x}}} = \dfrac{-1.2857 - 0}{0.5389} = -2.386 > -2.650$ (or $\overline{x} = -1.2857 > -1.4281$), we do not reject H_0. The new formula is not significantly more effective.

9-48 $\quad s_M^2 = 2.5714 \qquad n_M = 8 \qquad \overline{x}_M = 3.5 \qquad s_S^2 = 1.9821 \qquad n_S = 8 \qquad \overline{x}_S = 5.625$

$H_0 : \mu_M = \mu_S$ $\qquad$ $H_1 : \mu_M < \mu_S$ $\qquad$ $\alpha = .025$

$$s_p = \sqrt{\frac{(n_M - 1)s_M^2 + (n_S - 1)s_S^2}{n_M + n_S - 2}} = \sqrt{\frac{7(2.5714) + 7(1.9821)}{14}} = 1.5089$$

The lower limit of the acceptance region is $t_L = -t_{14,\ .025} = -2.145$, or

$$(\overline{x}_M - \overline{x}_S)_{CRIT} = 0 - t_{14,\ .025}s_p\sqrt{\frac{1}{n_M} + \frac{1}{n_S}} = -2.145(1.5089)\sqrt{\frac{1}{8} + \frac{1}{8}} = -1.6183$$

Since $t = \dfrac{(\overline{x}_M - \overline{x}_S) - (\overline{\mu}_M - \overline{\mu}_S)}{s_p\sqrt{\frac{1}{n_M} + \frac{1}{n_S}}} = \dfrac{(3.5 - 5.625) - 0}{1.5089\sqrt{\frac{1}{8} + \frac{1}{8}}} = -2.817 < -2.145$ (or $\overline{x}_M - \overline{x}_S = 3.5 - 5.625 = -2.125 < -1.6183$, we reject H_0. Severe consequences lead to a significantly greater attribution of responsibility.

9-50 $\quad$ Disney: $\quad n = 5 \qquad \sum x = 143.8 \qquad \sum x^2 = 4898.06$

$$\overline{x}_D = \frac{\sum x}{n} = \frac{143.8}{5} = 28.76$$

$$s_D^2 = \frac{1}{n-1}\left(\sum x^2 - n\overline{x}^2\right) = \frac{1}{4}\left(4898.06 - 5(28.76)^2\right) = 190.593$$

Competition: $n = 11$ $\quad \sum x = 156.7$ $\quad \sum x^2 = 3969.05$

$$\bar{x}_C = \frac{\sum x}{n} = \frac{156.7}{11} = 14.2455$$

$$s_C^2 = \frac{1}{10}\left(3969.05 - 11(14.2455)^2\right) = 173.6773$$

$H_0 : \mu_D = \mu_C$ $\quad H_1 : \mu_D > \mu_C$ $\quad \alpha = .05$

$$s_p = \sqrt{\frac{(n_D - 1)s_D^2 + (n_C - 1)s_C^2}{n_D + n_C - 2}} = \sqrt{\frac{4(190.593) + 10(173.6773)}{14}} = 13.3608$$

The upper limit of the acceptance region is $t_U = t_{14,\ .05} = 1.761$, or

$$(\bar{x}_D - \bar{x}_C)_U = (\mu_D - \mu_C)_{H_0} + t_{14,.05} s_p \sqrt{\frac{1}{n_D} + \frac{1}{n_C}} = 0 + 1.761(13.361)\sqrt{\frac{1}{5} + \frac{1}{11}} = 12.6903$$

Since $t = \dfrac{(\bar{x}_D - \bar{x}_C) - (\bar{\mu}_D - \bar{\mu}_C)}{s_p\sqrt{\frac{1}{n_D} + \frac{1}{n_C}}} = \dfrac{(28.76 - 14.2455) - 0}{13.361\sqrt{\frac{1}{5} + \frac{1}{11}}} = 2.014 > 1.761$ (or $\bar{x}_D - \bar{x}_C = 14.5145 > 12.6903$), we reject H_0. The Disney films earn significantly more than the competitors' films earn.

9-52 1995: $n_{95} = 2000$ $\quad \bar{p}_{95} = \dfrac{58}{2000} = 0.029$

1994: $n_{94} = 2500$ $\quad \bar{p}_{94} = \dfrac{61}{2500} = 0.0244$

$H_0 : p_{95} = p_{94}$ $\quad H_1 : p_{95} \neq p_{94}$ $\quad \alpha = .01$

$$\hat{p} = \frac{n_{95}\bar{p}_{95} + n_{94}\bar{p}_{94}}{n_{95} + n_{94}} = \frac{2000(.029) + 2500(.0244)}{2000 + 2500} = 0.0264$$

$$\hat{\sigma}_{\bar{p}_{95} - \bar{p}_{94}} = \sqrt{\hat{p}\hat{q}\left(\frac{1}{n_{95}} + \frac{1}{n_{94}}\right)} = \sqrt{.0264(.9736)\left(\frac{1}{2000} + \frac{1}{2500}\right)} = 0.0048$$

The limits of the acceptance region are $z_{CRIT} = z_{.005} = \pm 2.576$, or

$$(\bar{p}_{95} - \bar{p}_{94})_{CRIT} = 0 \pm z_{.01}\hat{\sigma}_{\bar{p}_{95} - \bar{p}_{94}} = 0 \pm 2.576(.0048) = \pm 0.0124$$

Since $z = \dfrac{\bar{p}_{95} - \bar{p}_{94}}{\hat{\sigma}_{\bar{p}_{95} - \bar{p}_{94}}} = \dfrac{.029 - .0244}{.00481} = 0.9563 < 2.576$ (or $\bar{p}_{95} - \bar{p}_{94} = 0.0046 < 0.0124$), we fail to reject H_0. The proportion of 1995 tax returns that were audited was not significantly different than the proportion of 1994 tax returns that were audited.

9-54 Differences between prices on 5/21/93 and 5/24/93 yield the following:
$s = 0.55232$, n= 20, $\bar{x} = -0.00625$

$$\hat{\sigma}_{\bar{x}} = s/\sqrt{n} = 0.55232/\sqrt{20} = 0.1235$$

$H_0 : \mu_{21} = \mu_{24}$ $\quad H_1 : \mu_{21} > \mu_{24}$ $\quad$ (α not given)

Since $t = \dfrac{\bar{x} - \mu}{\hat{\sigma}_{\bar{x}}} = \dfrac{-0.00625 - 0}{0.1235} = -0.0506$ is so close to 0, we would not reject H_0 at any reasonable significance level. The observed decrease is not significant.

9-56 $n_C = 280$ $\quad \bar{p}_C = \dfrac{152}{280} = 0.5429$ $\quad n_D = 190$ $\quad \bar{p}_D = \dfrac{81}{190} = 0.4263$

$H_0 : \bar{p}_C = \bar{p}_D$ $\quad H_1 : \bar{p}_C > \bar{p}_D$ $\quad \alpha = 0.02$

$$\hat{p} = \frac{n_C\overline{p}_C + n_D\overline{p}_D}{n_C + n_D} = \frac{280(.5429) + 190(.4263)}{280 + 190} = 0.4958$$

$$\hat{\sigma}_{\overline{p}_C - \overline{p}_D} = \sqrt{\hat{p}\hat{q}\left(\frac{1}{n_C} + \frac{1}{n_D}\right)} = \sqrt{.4958(.5042)\left(\frac{1}{280} + \frac{1}{190}\right)} = 0.047$$

The upper limit of the acceptance region is $z_U = z_{.02} = 2.05$, or

$$(\overline{p}_C - \overline{p}_D)_U = 0 + z_{.02}\hat{\sigma}_{\overline{p}_C - \overline{p}_D} = 0 + 2.05(.047) = 0.0964$$

Since $z = \dfrac{\overline{p}_C - \overline{p}_D}{\hat{\sigma}_{\overline{p}_C - \overline{p}_D}} = \dfrac{.5429 - .4263}{.0470} = 2.48 > 2.05$ (or $\overline{p}_C - \overline{p}_D = 0.1166 > 0.047$), we reject H_0. Cat owners are significantly more likely to feed their pets premium food than dog owners.

9-58 a) $59(.45) = 25.55$ responses, which is impossible!
Note that $26/59 = .4407$, but $27/59 = .4576$.

b) Looking at the greatest difference in response rates (with 26 U.K. responses), we have

$n_{US} = 100 \quad \overline{p}_{US} = \frac{50}{100} = 00.5 \quad n_{UK} = 59 \quad \overline{p}_{UK} = \frac{26}{59} = .4407$

$H_0: \overline{p}_{US} = \overline{p}_{UK} \quad H_1: \overline{p}_{US} \neq \overline{p}_{UK}$ no α is given

$$\hat{p} = \frac{n_{US}\overline{p}_{US} + n_{UK}\overline{p}_{UK}}{n_{US} + n_{UK}} = \frac{100(0.5) + 59(.4407)}{100 + 59} = 0.4780$$

$$\hat{\sigma}_{\overline{p}_{US} - \overline{p}_{UK}} = \sqrt{\hat{p}\hat{q}\left(\frac{1}{n_{US}} + \frac{1}{n_{UK}}\right)} = \sqrt{.4780(.5220)\left(\frac{1}{100} + \frac{1}{59}\right)} = 0.0820$$

Now $z = \dfrac{\overline{p}_{US} - \overline{p}_{UK}}{\hat{\sigma}_{\overline{p}_{US} - \overline{p}_{UK}}} = \dfrac{0.5 - .4407}{.0820} = 0.7232$, which corresponds to a probability value of $2(0.5 - 0.2642) = 0.4716$, which is very insignificant. The proportions of responses in the two surveys are not significantly different.

CHAPTER 10

QUALITY AND QUALITY CONTROL

10-2 As in 10-1, there are many potential answers to this question. Any example that demonstrates how a low cost item works reliably or lacks defects would be appropriate. Examples include paper clips, shoe laces, toothpicks, coins, or pencils.

10-4 Quality control is an important issue to management because good managers want to keep their customers satisifed. From another perspective, quality impacts the "bottom line" -- profits. Parts that don't conform to requirements end up as scrap or rejects, which is costly. It is also costly to inspect for defects. By controlling the quality of the parts used in production at all stages of the process, costs can be minimized.

10-6 "Zero defects" is a goal in the production process, where defects are eliminated at each stage of the production, so that the final product is a perfectly reliable, quality product.

10-8 Mechanical devices, such as a robot, are created with a specific and limited range of motion. Materials and parts used to build such machines are chosen to insure uniformity and reliability. Robots are therefore, easily controlled and adjusted. In contrast, humans are capable of a great variety of motions and react to all sorts of stimuli (both related and unrelated to the task at hand) which affects behavior. It is expected that for those reasons, humans would introduce more random variation into the work process.

10-10 The manager would be trying to control the amount of time to process each customer. By directing customers with just a few items to the express line, you would in effect increase the consistency in the times taken to process each customer, since the registers would be handling customers with a similar number of items. Without express lines, one particular register would process customers with greatly varying numbers of items which would cause greater variation in processing time.

10-12 a) $\overline{\overline{x}} = 16.4$ $\sigma_{\overline{x}} = 1.2$

$$\text{CL} = \overline{\overline{x}} = 16.4$$
$$\text{UCL} = \overline{\overline{x}} + 3\sigma_{\overline{x}} = 16.4 + 3(1.2) = 20.0$$
$$\text{LCL} = \overline{\overline{x}} - 3\sigma_{\overline{x}} = 16.4 - 3(1.2) = 12.8$$

b) $\overline{\overline{x}} = 16.4$ $\overline{R} = 7.6$ $n = 12$ $d_2 = 3.258$

$$\text{CL} = \overline{\overline{x}} = 16.4$$
$$\text{UCL} = \overline{\overline{x}} + \frac{3(\overline{R})}{d_2\sqrt{n}} = 16.4 + \frac{3(7.6)}{3.258\sqrt{12}} = 18.42$$
$$\text{LCL} = \overline{\overline{x}} - \frac{3(\overline{R})}{d_2\sqrt{n}} = 16.4 - \frac{3(7.6)}{3.258\sqrt{12}} = 14.38$$

c) $\overline{\overline{x}} = 4.1$ $\overline{R} = 1.3$ $n = 8$ $d_2 = 2.847$

$$\text{CL} = \overline{\overline{x}} = 4.1$$
$$\text{UCL} = \overline{\overline{x}} + \frac{3(\overline{R})}{d_2\sqrt{n}} = 4.1 + \frac{3(1.3)}{2.847\sqrt{8}} = 4.58$$
$$\text{LCL} = \overline{\overline{x}} - \frac{3(\overline{R})}{d_2\sqrt{n}} = 4.1 - \frac{3(1.3)}{2.847\sqrt{8}} = 3.62$$

d) $\overline{\overline{x}} = 141.7$ $\overline{R} = 18.6$ $n = 15$ $d_2 = 3.472$

$\text{CL} = \overline{\overline{x}} = 141.7$

$\text{UCL} = \overline{\overline{x}} + \dfrac{3(\overline{R})}{d_2\sqrt{n}} = 141.7 + \dfrac{3(18.6)}{3.472\sqrt{15}} = 145.85$

$\text{LCL} = \overline{\overline{x}} - \dfrac{3(\overline{R})}{d_2\sqrt{n}} = 141.7 - \dfrac{3(18.6)}{3.472\sqrt{15}} = 137.55$

10-14 a) $n = 9$ $k = 21$ $A_2 = 0.337$

$\overline{\overline{x}} = \dfrac{\sum \overline{x}}{k} = \dfrac{312.999}{21} = 14.905$ $\overline{R} = \dfrac{\sum R}{k} = \dfrac{385.300}{21} = 18.348$

$\text{CL} = \overline{\overline{x}} = 14.905$

$\text{UCL} = \overline{\overline{x}} + A_2(\overline{R}) = 14.905 + 0.337(18.348) = 21.09$

$\text{LCL} = \overline{\overline{x}} - A_2(\overline{R}) = 14.905 - 0.377(18.348) = 8.72$

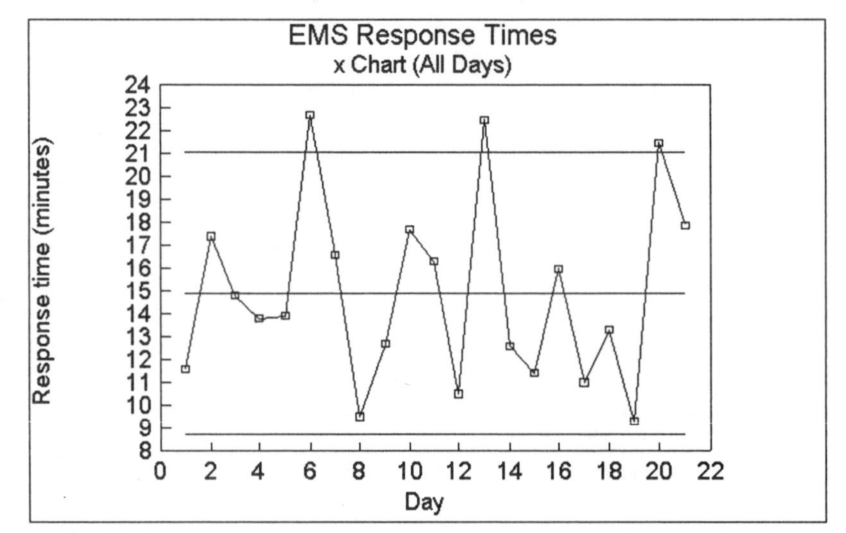

b) It seems that response times are out of control (outliers on the high side) on Saturdays. Dick should investigate whether there are any special circumstances that tend to repeat on Saturdays. For example, there might be more calls coming in on Saturdays which burden the capabilities of the rescue squads. To counteract this problem, Dick might consider increasing the number of squads on call on Saturdays, or he might provide additional training to the crews on Saturdays.

c) No Saturdays:

$\overline{\overline{x}} = \dfrac{\sum \overline{x}}{k} = \dfrac{287.343}{21} = 13.683$ $\overline{R} = \dfrac{\sum R}{k} = \dfrac{396.600}{21} = 17.6$

$\text{CL} = \overline{\overline{x}} = 13.683$

$\text{UCL} = \overline{\overline{x}} + A_2(\overline{R}) = 13.683 + 0.337(17.6) = 19.61$

$\text{LCL} = \overline{\overline{x}} - A_2(\overline{R}) = 13.683 - 0.337(17.6) = 7.75$

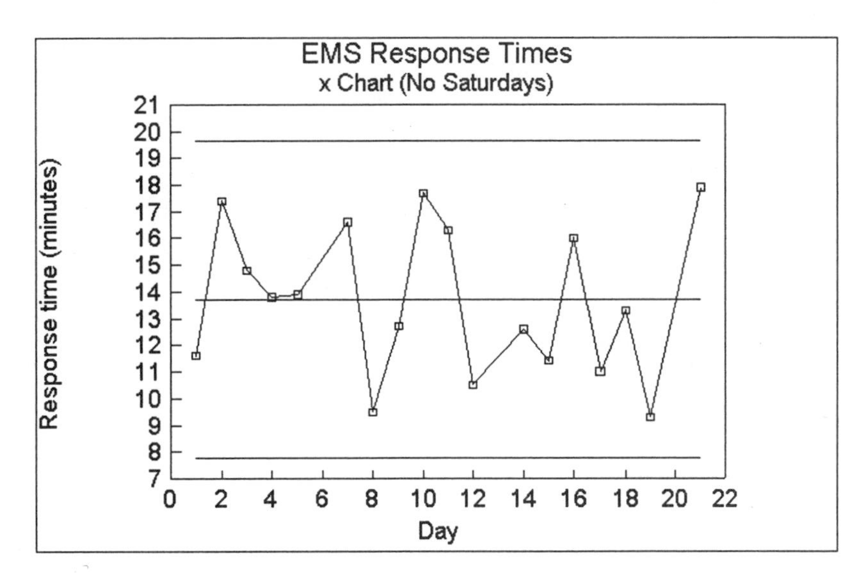

Although some variability remains, the response times are within the control limits when data for Saturdays are removed.

10-16 a) $n = 15$ $k = 24$ $A_2 = 0.233$

$$\overline{\overline{x}} = \frac{\sum \overline{x}}{k} = \frac{96.5208}{24} = 4.0217 \qquad \overline{R} = \frac{\sum R}{k} = \frac{2.4192}{24} = 0.1008$$

$\text{CL} = \overline{\overline{x}} = 4.0217$

$\text{UCL} = \overline{\overline{x}} + A_2(\overline{R}) = 4.0217 + 0.233(0.1008) = 4.045$

$\text{LCL} = \overline{\overline{x}} - A_2(\overline{R}) = 4.0217 - 0.233(0.1008) = 3.998$

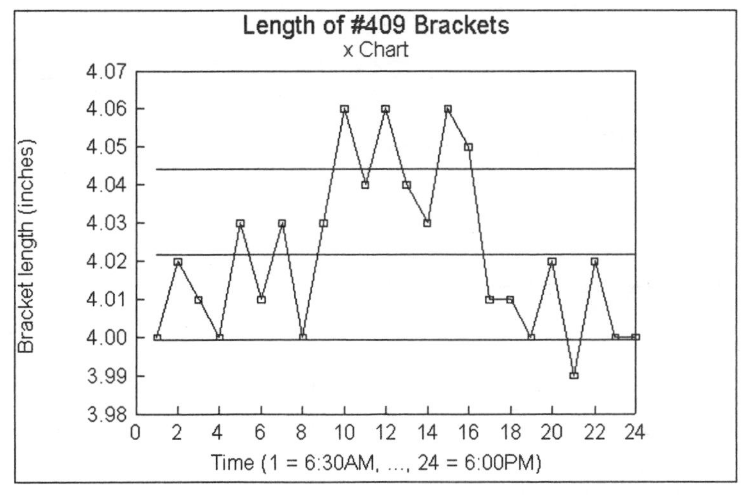

b) The three shifts seem to be at different levels, with the bracket lengths in the second shift higher than in the first and third shifts. Silvia should check out the procedures for recalibrating the saw at the beginning of each shift

10-18 Since new apprentices are expected to make errors, there should be greater variability in the spindles they produce initially. This variability should decrease as the new apprentice gains experience. Pattern (a) conforms to this expectation.

10-20 a) $n = 9$ $k = 21$ $D_4 = 1.816$ $D_3 = 0.184$

$$\overline{R} = \frac{\sum R}{k} = \frac{\sum 385.3}{21} = 18.348$$

$\text{CL} = \overline{R} = 18.348$
$\text{UCL} = \overline{R}\,\text{D}_4 = 18.348(1.816) = 33.32$
$\text{LCL} = \overline{R}\,\text{D}_3 = 18.348(0.184) = 3.38$

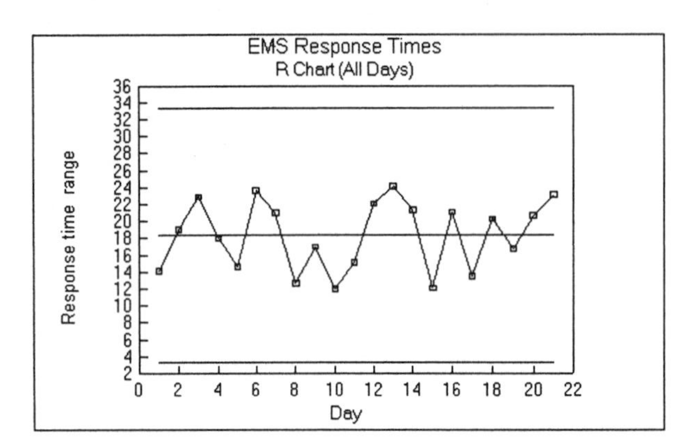

b) The Saturdays are no longer outliers, but they do tend to have higher variability than the other days. This could well arise because of the greater number of calls coming in on Saturdays.

c) No Saturdays:

$n = 9 \qquad k = 18 \qquad \text{D}_4 = 1.816 \qquad \text{D}_3 = 0.184$

$$\overline{R} = \frac{\sum R}{k} = \frac{\sum 316.8}{18} = 17.6$$

$\text{CL} = \overline{R} = 17.6$

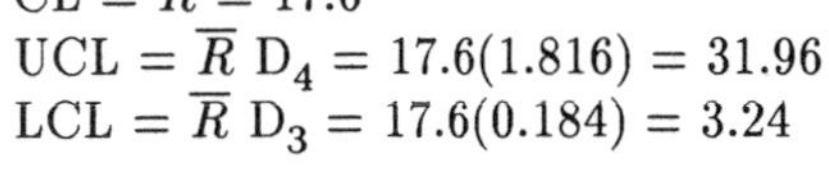

$\text{UCL} = \overline{R}\,\text{D}_4 = 17.6(1.816) = 31.96$
$\text{LCL} = \overline{R}\,\text{D}_3 = 17.6(0.184) = 3.24$

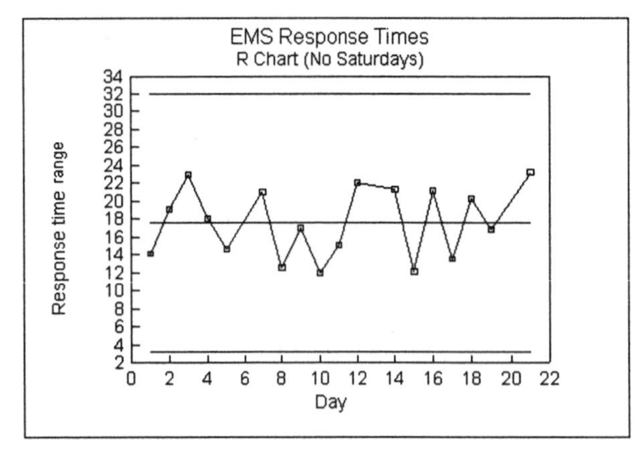

The variability in response times is in-control.

10-22 $\quad n = 15 \qquad k = 24 \qquad \text{D}_4 = 1.653 \qquad \text{D}_3 = 0.347 \qquad \overline{R} = 0.1008$
$\text{CL} = \overline{R} = 0.1008 \qquad \text{UCL} = \overline{R}\,\text{D}_4 = 0.1008(1.653) = 0.167$
$\text{LCL} = \overline{R}\,\text{D}_3 = 0.1008(0.347) = 0.035$

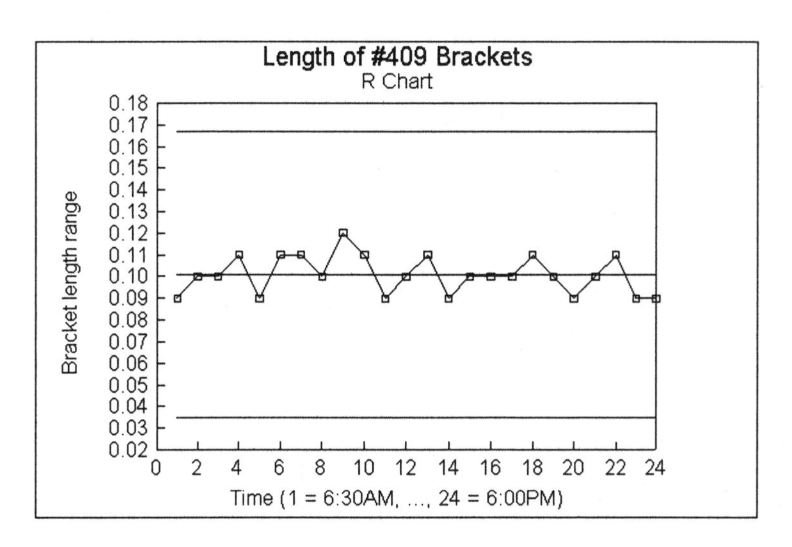

The variability in the process appears to be well in control.

10-24 a) $\text{CL} = \bar{\bar{p}} = 0.25$

$$\text{UCL} = \bar{\bar{p}} + 3\sqrt{\frac{pq}{n}} = 0.25 + 3\sqrt{\frac{.25(.75)}{30}} = 0.487$$

$$\text{LCL} = \bar{\bar{p}} - 3\sqrt{\frac{pq}{n}} = 0.25 - 3\sqrt{\frac{.25(.75)}{30}} = 0.013$$

b) $\text{CL} = \bar{\bar{p}} = 0.15$

$$\text{UCL} = \bar{\bar{p}} + 3\sqrt{\frac{pq}{n}} = 0.15 + 3\sqrt{\frac{.15(.85)}{65}} = 0.283$$

$$\text{LCL} = \bar{\bar{p}} - 3\sqrt{\frac{pq}{n}} = 0.15 - 3\sqrt{\frac{.15(.85)}{65}} = 0.017$$

c) $\text{CL} = \bar{\bar{p}} = 0.05$

$$\text{UCL} = p + 3\sqrt{\frac{pq}{n}} = 0.05 + 3\sqrt{\frac{.05(.95)}{82}} = 0.122$$

$$\text{LCL} = p - 3\sqrt{\frac{pq}{n}} = 0.05 - 3\sqrt{\frac{.05(.95)}{82}} < 0, \text{ so the LCL} = 0$$

d) $\text{CL} = p = 0.42$

$$\text{UCL} = p + 3\sqrt{\frac{pq}{n}} = 0.42 + 3\sqrt{\frac{.42(.58)}{97}} = 0.570$$

$$\text{LCL} = p - 3\sqrt{\frac{pq}{n}} = 0.42 - 3\sqrt{\frac{.42(.58)}{97}} = 0.270$$

e) $\text{CL} = \bar{\bar{p}} = 0.63$

$$\text{UCL} = \bar{\bar{p}} + 3\sqrt{\frac{pq}{n}} = 0.63 + 3\sqrt{\frac{.63(.37)}{124}} = 0.760$$

$$\text{LCL} = \bar{\bar{p}} - 3\sqrt{\frac{pq}{n}} = 0.63 - 3\sqrt{\frac{.63(.37)}{124}} = 0.500$$

10-26 a) H_0: $p = 0.015$ H_1: $p > 0.015$

$p = 0.01594$ $$z = \frac{\bar{p} - \mu_{\bar{p}}}{\sqrt{\frac{pq}{n}}} = \frac{.01594 - .015}{\sqrt{\frac{.015(.985)}{16000}}} = 0.98$$

Since the probability value = 0.5 − 0.3365 = 0.1635, we accept the H_0. She can be reasonably sure that the proportion of bad capsules is not significantly greater than 1.5%.

b) $n = 500 \qquad p = 0.015$

$\text{CL} = p = 0.015$

$$\text{UCL} = p + 3\sqrt{\frac{pq}{n}} = 0.015 + 3\sqrt{\frac{.015(.985)}{500}} = 0.0313$$

$$\text{LCL} = p - 3\sqrt{\frac{pq}{n}} = 0.015 - 3\sqrt{\frac{.898(.102)}{150}} = 0$$

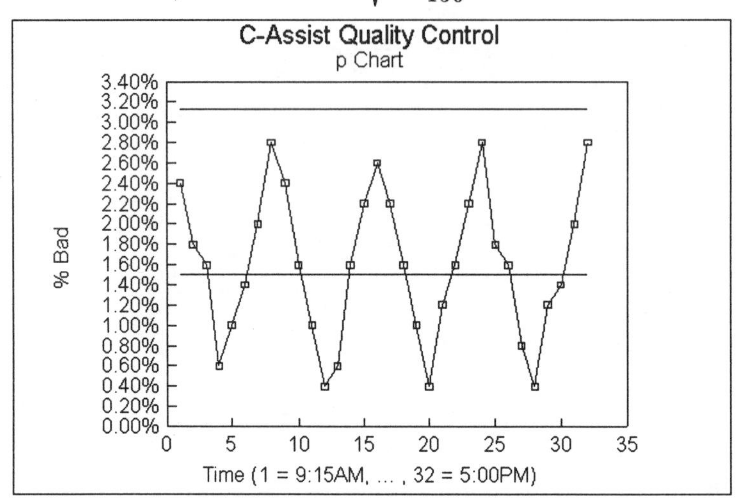

c) Although, the samples fall within the control limits, the p chart shows a distinct 2-hour cycle in which the samples of capsules approach the upper control limit. Sherry needs to examine the manufacturing process to determine what is causing the cycling.

10-28 $n = 240 \qquad k = 30 \qquad \overline{\overline{p}} = \frac{\sum \overline{p}}{k} = \frac{2.043}{30} = 0.0681$

$\text{CL} = \overline{\overline{p}} = 0.0681$

$$\text{UCL} = \overline{\overline{p}} + 3\sqrt{\frac{\text{pq}}{\text{n}}} = 0.0681 + 3\sqrt{\frac{.0681(.9319)}{240}} = 0.1169$$

$$\text{LCL} = \overline{\overline{p}} - 3\sqrt{\frac{\text{pq}}{\text{n}}} = 0.0681 - 3\sqrt{\frac{.0681(.9319)}{240}} = 0.0193$$

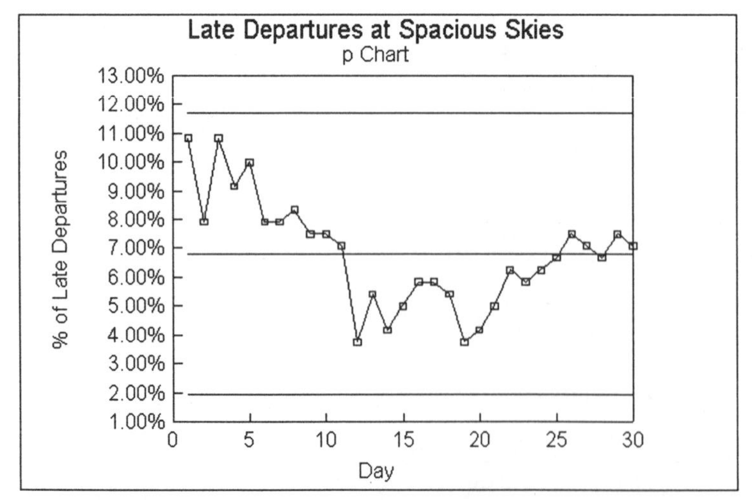

Four weeks ago, the fraction of late departures dropped dramatically, presumably in response to Ross' new procedures. However, in the past two weeks, that fraction has again started drifting upward. If the new procedures aren't being used, he should insist that they be used. If they are being used, he needs to find out why they aren't working.

10-30 After the most important problems are "slain", other problems will likely become "dragons", especially if they are ignored. It should be stressed to Joe, that for TQM to succeed, there must be a focus on "continuous quality improvement."

10-32

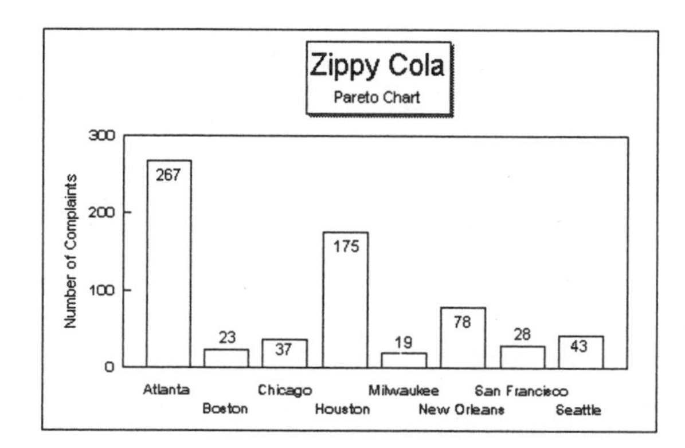

The plants in Atlanta and Houston should be visited first.

10-34 Total inspection is frequently impractical because of the time and cost involved.

10-36 AQL = 0.02. Numerical results obtained from an Excel spreadsheet.

a) $n = 175,\ c = 3$: $P(r \geq 4) = 0.46446$
b) $n = 175,\ c = 5$: $P(r \geq 6) = 0.14037$
c) $n = 250,\ c = 3$: $P(r \geq 4) = 0.73781$
d) $n = 250,\ c = 5$: $P(r \geq 6) = 0.38403$

10-38 a) AQL = 0.005
$p = 0.5\%$
The probability of acceptance (Y axis) $\approx$ 0.87
Therefore, the Producer's Risk $\approx 1 - .87 = 0.13$

b) AQL = 0.010
$p = 1.0\%$
The probability of acceptance (Y axis) $\approx$ 0.54
Therefore, the Producer's Risk $\approx 1 - .54 = 0.46$

c) AQL = 0.015
$p = 1.5\%$
The probability of acceptance (Y axis) $\approx$ 0.28
Therefore, the Producer's Risk $\approx 1 - .28 = 0.72$

10-40 a) H_0: $p = 0.02$ H_1: $p > 0.02$

$p = 0.0225$

$$z = \frac{\overline{p} - \mu_{\overline{p}}}{\sqrt{\frac{pq}{n}}} = \frac{.0225 - .02}{\sqrt{\frac{.02(.98)}{2000}}} = 0.80$$

Since the probability value = 0.5 − 0.2881 = 0.2119, we accept the H_0. She can be reasonably sure that the proportion of audited clients is not significantly greater than 2.0%.

b) $n = 125$ $p = 0.02$
CL = p = 0.02

$$UCL = p + 3\sqrt{\frac{pq}{n}} = 0.02 + 3\sqrt{\frac{.02(.98)}{125}} = 0.0576$$

$$LCL = p - 3\sqrt{\frac{pq}{n}} = 0.02 - 3\sqrt{\frac{.02(.98)}{125}} = 0$$

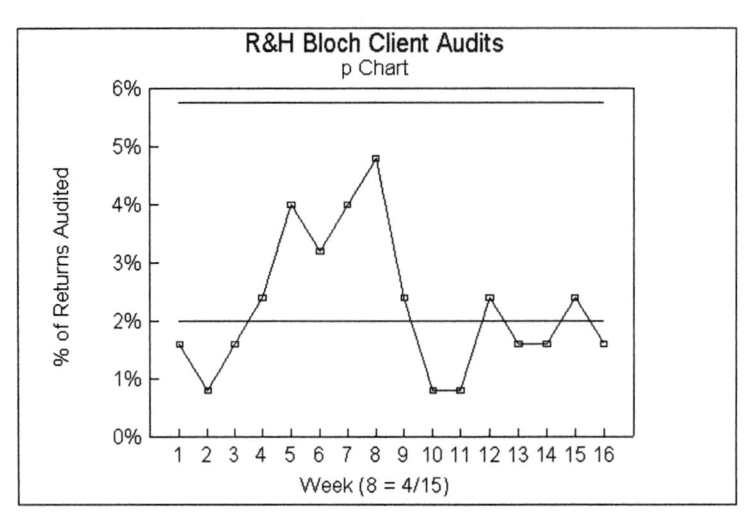

Although, the samples fall within the control limits, the p chart indicates that the percent audited has taken a jump upwards in the four weeks prior to April 15. This may indicate something about the clients who wait until the last minute, or something about how the IRS chooses which returns to audit. In either case, the partners should be aware of this phenomenon.

10-42 Attributes are categorical variables with only two possible categories.

10-44 a) $n = 10$ $\quad k = 16$ $\quad A_2 = 0.308$

$$\overline{\overline{x}} = \frac{\sum \overline{x}}{k} = \frac{800}{16} = 50 \qquad \overline{R} = \frac{\sum R}{k} = \frac{108}{16} = 6.75$$

$$\text{CL} = \overline{\overline{x}} = 50$$

$$\text{UCL} = \overline{\overline{x}} + A_2(\overline{R}) = 50 + 0.308(6.75) = 52.08$$

$$\text{LCL} = \overline{\overline{x}} - A_2(\overline{R}) = 50 - 0.308(6.75) = 47.92$$

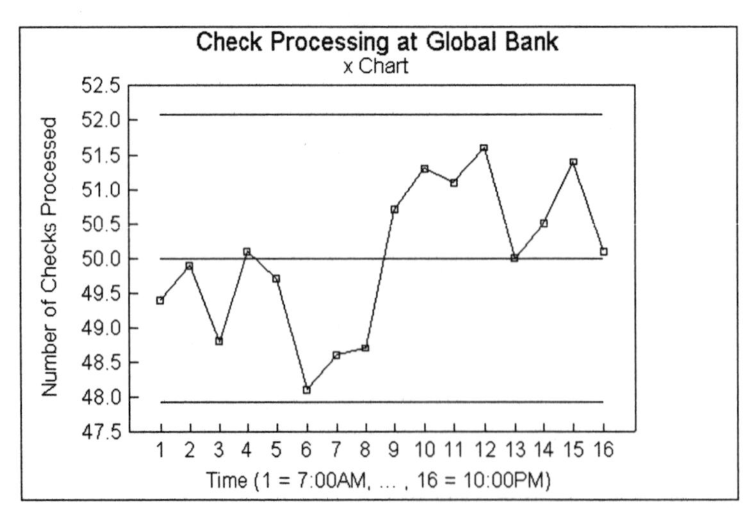

b) The output during the second shift is at a higher level. Shih-Hsing should try to learn why productivity is higher on that shift.

10-46

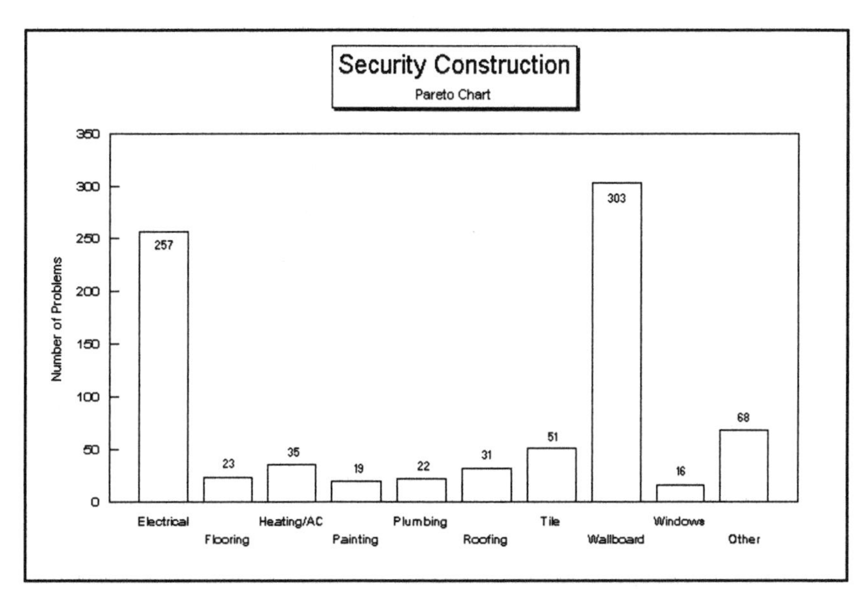

The wallboard and electrical subcontractors will require additional supervision.

10-48 LTPD = 0.015

a) $n = 200 \qquad c = 1$

$r = 0$: $\frac{n!}{r!(n-r)!}\,p^r q^{n-r} = \frac{200!}{0!(200)!}\,(.015)^0(.985)^{200} = 1(1)(.0487) = 0.0487$

$r = 1$: $\frac{n!}{r!(n-r)!}\,p^r q^{n-r} = \frac{200!}{1!(199)!}\,(.015)^1(.985)^{199} = 200(.015)(.0494) = 0.1482$

$.0487 + .1482 = 0.1969$

b) $n = 200 \qquad c = 2$

$r = 2$: $\frac{n!}{r!(n-r)!}\,p^r q^{n-r} = \frac{200!}{2!(198)!}\,(.015)^2(.985)^{198} = 19900(.000225)(.0502) = 0.2246$

$.0487 + .1482 + .2246 = 0.4215$

c) $n = 250 \qquad c = 1$

$r = 0$: $\frac{n!}{r!(n-r)!}\,p^r q^{n-r} = \frac{250!}{0!(250)!}\,(.015)^0(.985)^{250} = 1(1)(.0229) = 0.0229$

$r = 1$: $\frac{n!}{r!(n-r)!}\,p^r q^{n-r} = \frac{250!}{1!(249)!}\,(.015)^1(.985)^{249} = 250(.015)(.0232) = 0.087$

$.0229 + .087 = 0.1099$

d) $n = 250 \qquad c = 2$

$r = 2$: $\frac{n!}{r!(n-r)!}\,p^r q^{n-r} = \frac{250!}{2!(248)!}\,(.015)^2(.985)^{248} = 31125(.000225)(.0236) = 0.1650$

$.0229 + .087 + .1650 = 0.2749$

10-50 a) $n = 24 \qquad k = 20 \qquad A_2 = 0.157$

$\overline{\overline{x}} = \frac{\sum \overline{x}}{k} = \frac{1499.3}{20} = 74.965 \qquad \overline{R} = \frac{\sum R}{k} = \frac{63.3}{20} = 3.165$

$\text{CL} = \overline{\overline{x}} = 74.965$

$\text{UCL} = \overline{\overline{x}} + A_2(\overline{R}) = 74.965 + 0.157(3.165) = 75.462$

$\text{LCL} = \overline{\overline{x}} - A_2(\overline{R}) = 74.965 - 0.157(3.165) = 74.468$

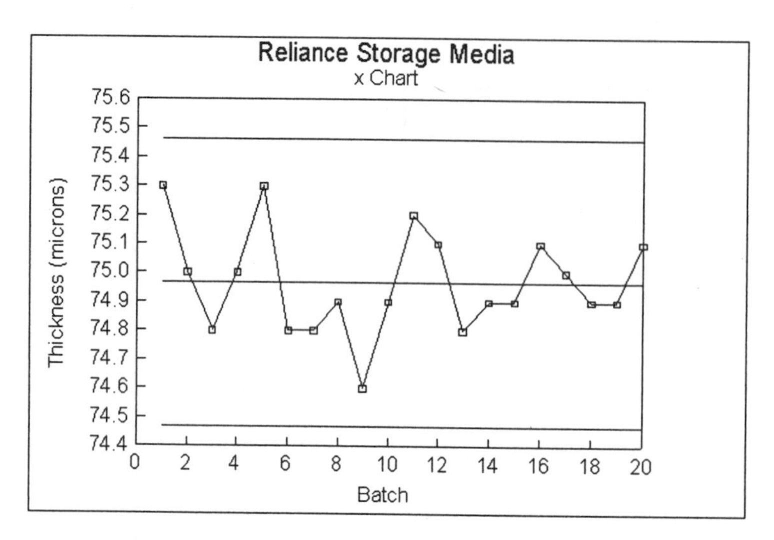

b) The process is in-control. There are no outliers or out-of-control patterns.

c) Yes. The last 10 observations do cluster closer to the center line than do the first 10 observations. Deshawn should be happy to see this pattern, since it indicates the inherent variability of the process has decreased. To the extent this is true, he might want to use the last 10 observations to re-compute the $\overline{x}$ chart. The new chart will have narrower control limits.

10-52 $n = 2000 \qquad p = 0.001$

CL $= p = 0.001$

$$\text{UCL} = p + 3\sqrt{\frac{pq}{n}} = 0.001 + 3\sqrt{\frac{.001(.999)}{2000}} = 0.0031$$

$$\text{LCL} = p - 3\sqrt{\frac{pq}{n}} = 0.001 - 3\sqrt{\frac{.001(.999)}{2000}} = 0$$

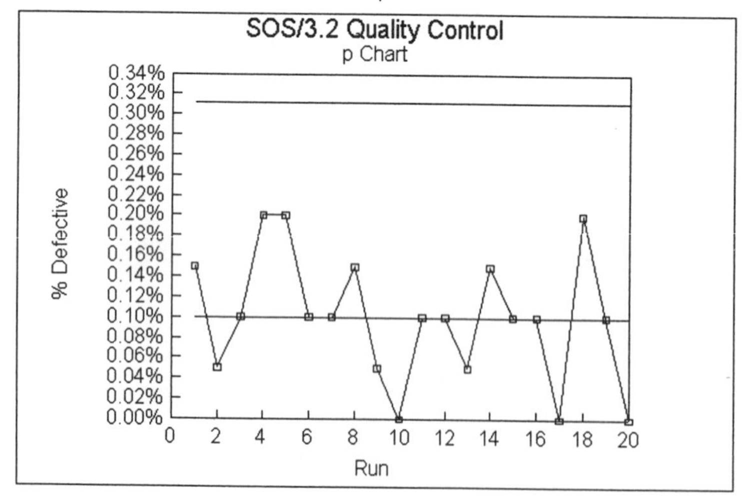

The chart shows the process is in-control.

10-54 Common variation: density of flour, variability in measuring ingredients, variability in the amount of dough per cracker.

Special cause variation: improper calibration of measuring machinery, drifting termperature in ovens, miscounting by packaging machinery.

10-56 a) LTPD = 0.010
$p = 1.0\%$
The probability of acceptance (Y axis) $\approx$ 0.64
Therefore, the Consumer's Risk $\approx$ 0.64

b) LTPD = 0.015
$p = 1.5\%$
The probability of acceptance (Y axis) $\approx$ 0.34
Therefore, the Consumer's Risk $\approx$ 0.34

c) LTPD = 0.020
$p = 2.0\%$
The probability of acceptance (Y axis) $\approx$ 0.15
Therefore, the Consumer's Risk $\approx$ 0.15

10-58 Acceptance sampling is more effective in the long run than sampling entire batches because it forces the supplier to take responsibility for the quality of the output.

10-60 a) No. This is a good example of <u>inspection</u> instead of <u>prevention.</u>

b) Major bones could collect causes by parent (failure to make appointment, failure to keep appointment, etc.), by child (illness at time of appointment, allergic reactions, etc.), and by health care professionals (shortage of vaccine, errors in record keeping, etc.).

c) Collecting data from a sample of missed immunizations would enable the HMO to construct a Pareto diagram to help identify the principal causes of the problem.

CHAPTER 11

CHI-SQUARE AND ANALYSIS OF VARIANCE

11-2 To determine if three or more population means can be considered equal

11-4 a) FALSE; can do inference only on one or two variances

b) TRUE; use analysis of variance

c) TRUE; use a chi-square test

11-6 $df = (\# \text{ rows} - 1)(\# \text{ columns} - 1)$

a) $4 \times 3 = 12$ b) $5 \times 1 = 5$ c) $2 \times 6 = 12$ d) $3 \times 3 = 9$

11-8 a)

f_o	f_e	$f_o - f_e$	$(f_o - f_e)^2$	$\frac{(f_o - f_e)^2}{f_e}$
12	20.2	−8.2	67.24	3.3287
18	20.2	−2.2	4.84	0.2396
17	20.2	−3.2	10.24	0.5069
22	20.2	1.8	3.24	0.1604
32	20.2	11.8	139.24	6.8931
18	25.2	−7.2	51.84	2.0571
25	25.2	−0.2	0.04	0.0016
29	25.2	3.8	14.44	0.5730
24	25.2	−1.2	1.44	0.0571
30	25.2	4.8	23.04	0.9143
45	29.6	15.4	237.16	8.0122
32	29.6	2.4	5.76	0.1946
29	29.6	−0.6	0.36	0.0122
29	29.6	−0.6	0.36	0.0122
13	29.6	−16.6	275.56	9.3095

$$\chi^2 = \sum \frac{(f_o - f_e)^2}{f_e} = 32.2725$$

b) H_0 : Age group is independent of purchasing plans
H_1 : Age group is related to purchasing plans

c) $df = (3-1) \times (5-1) = 8$ $\chi^2_{CRIT} = \chi^2_{.01;8} = 20.090$

Thus we reject H_0 because $\chi^2 > \chi^2_{CRIT}$, and we conclude there is a relationship between age and purchasing plans.

11-10 a) H_0 : weekly chip sales independent of economy
H_1 : weekly chip sales related to state of economy

b)

f_o	f_e	$f_o - f_e$	$(f_o - f_e)^2$	$\frac{(f_o - f_e)^2}{f_e}$
20	15	5	25	1.667
7	9	−2	4	0.444
3	6	−3	9	1.500
30	50	−20	400	8.000
40	30	10	100	3.333
30	20	10	100	5.000
20	15	5	25	1.667
8	9	−1	1	0.111
2	6	−4	16	2.667
30	20	10	100	5.000
5	12	−7	49	4.083
5	8	−3	9	1.125

$$\chi^2 = \sum \frac{(f_o - f_e)^2}{f_e} = 34.597$$

c) $df = 2 \times 3 = 6$ $\qquad \chi^2_{CRIT} = \chi^2_{.10;6} = 10.645$

reject $H_0 \Rightarrow$ silicon chip sales are not independent of U.S. economy

11-12 H_0 : frequency of readership and level of education are independent
H_1 : they are not independent

	Prof	College	HS	< HS
Never				
Observed	10.000	17.000	11.000	21.000
Expected	18.643	21.275	8.993	10.089
Sometimes				
Observed	12.000	23.000	8.000	5.000
Expected	15.167	17.309	7.316	8.208
AM or PM				
Observed	35.000	38.000	16.000	7.000
Expected	30.335	34.617	14.632	16.416
Both editions				
Observed	28.000	19.000	6.000	13.000
Expected	20.855	23.799	10.059	11.286

f_o	f_e	$f_o - f_e$	$(f_o - f_e)^2$	$\frac{(f_o - f_e)^2}{f_e}$
10	18.643	- −8.643	74.701	4.007
17	21.275	−4.275	18.276	0.859
11	8.993	2.007	4.028	0.448
21	10.089	10.911	119.050	11.800
12	15.167	−3.167	10.030	0.661
23	17.309	5.691	32.387	1.871
8	7.316	0.684	0.468	0.064
5	8.208	−3.208	10.291	1.254
35	30.335	4.665	21.762	0.717
38	34.617	3.383	11.445	0.331
16	14.632	1.368	1.871	0.128
7	16.416	−9.416	88.661	5.401
28	20.855	7.145	51.051	2.448

19	23.799	−4.799	23.030	0.968
6	10.059	−4.059	16.475	1.638
13	11.286	1.714	2.938	0.260

$$\chi^2 = \sum \frac{(f_o - f_e)^2}{f_e} = 32.855$$

$df = 3 \times 3 = 9$ $\chi^2_{CRIT} = \chi^2_{.10;9} = 14.684$

reject $H_0 \Rightarrow$ frequency of readership differs according to education

11-14 a)

Sales(x)	2.6	3.8	5	6.2	7.4
$z = \frac{x-5}{1.5}$	−1.6	−0.8	0	0.8	1.6

Probability .0548 .1571 .2881 .2881 .1571 .0548

b)

Value	< 2.60	2.60-3.79	3.80-4.99	5.0-6.19	6.20-7.39	≥ 7.40
Observed	6	30	41	52	12	9
Expected	8.220	23.565	43.215	43.215	23.565	8.220
$\frac{(f_o-f_e)^2}{f_e}$	0.600	1.757	0.114	1.786	5.676	0.074

c) $\chi^2 = 0.600 + 1.757 + 0.114 + 1.786 + 5.676 + 0.074 = 10.007$

d) $df = 6 - 1 = 5$ $\chi^2_{CRIT} = \chi^2_{.10;5} = 9.236$

reject $H_0 \Rightarrow$ no, the distribution is not well described as normal with $\mu = 5$ and $\sigma = 1.5$

11-16 Let x = sales by Mr. Armstrong
H_0 : x distributed binomial (n=5, p=.4)
H_1 : x not distributed as above (it could be binomial with a different p or not even binomial at all)
$\alpha = .05$, $df = 4$ after pooling together the last two groups

# of sales/ day	0	1	2	3	4	5	4 or 5
Observed	10	41	60	20	6	3	9
Binomial prob.	.0778	.2592	.3456	.2304	.0768	.0102	.0870
Expected	10.89	36.29	48.38	32.26	10.75	1.43	12.18
$\frac{(f_o - f_e)^2}{f_e}$	.073	.611	2.791	4.659	combine because $f_e < 5$		.830

$$\chi^2 = \sum \frac{(f_o - f_e)^2}{f_e} = 8.964$$

$\chi^2_{CRIT} = \chi^2_{.05;4} = 9.488$, so do not reject $H_0 \Rightarrow$ Mr. Armstrong's sales may be described as binomially distributed, with $n = 5$, $p = .4$

11-18 a)

Dollars(x)	1000	2000
$z = \frac{x-1500}{600}$	−0.83	0.83

Deposits	0-999	1000-1999	2000 and more
Observed	20	65	25
Normal prob.	.2033	.5934	.2033
Expected	22.363	65.274	22.363
$\frac{(f_o - f_e)^2}{f_e}$	.2497	.0012	.3109

b) $\chi^2 = .2497 + .0012 + .3109 = .5618$

c) H_0 : deposits are normally distributed with $\mu = 1500$, $\sigma = 600$
H_0 : deposits are not distributed as above

d) $\chi^2_{CRIT} = \chi^2_{.10;2} = 4.605 \Rightarrow$ do not reject $H_0 \Rightarrow$ deposits are normally distributed with $\mu = 1500$ and $\sigma = 600$

11-20 H_0 : the winnings are distributed as stated $\qquad H_1$: winning probabilities are different

Winnings	\$1	\$100	\$0
Probability	.1	.05	.85
Observed	87	48	865
Expected	100	50	850
$\frac{(f_o - f_e)^2}{f_e}$	1.69	0.08	.2647

$\chi^2 = 1.69 + 0.08 + 0.2647 = 2.0347 \qquad \chi^2_{CRIT} = \chi^2_{.05;2} = 5.991$

do not reject $H_0 \Rightarrow$ the state's claim is reasonable

11-22 H_0 : binomial with $p = .3$, $n = 3 \qquad H_1$: some other distribution
Test at $\alpha = .05$ with 2 df

# of shifts	0	1	2 or 3
Binomial prob.	.343	.441	.216
Observed	16	27	17
Expected	20.58	26.46	12.96
$\frac{(f_o - f_e)^2}{f_e}$	1.019	0.011	1.259

$\chi^2 = \sum \frac{(f_o - f_e)^2}{f_e} = 2.289 \qquad \chi^2_{CRIT} = \chi^2_{.05;2} = 5.991$

accept $H_0 \Rightarrow$ the number of alarms is well described by a binomial distribution with $p = .3$ and $n = 3$.

11-24 H_0 : the number of customers arriving in 5-minute intervals is distributed Poisson with $\lambda = 3$
H_1 : not H_0
Test at $\alpha = .05$ with $df = 7 - 1 = 6$

# of customers	0	1	2	3	4	5	≥ 6
Poisson prob.	.0498	.1494	.2240	.2240	.1680	.1008	.0840
Observed	22	74	115	95	94	80	20
Expected	24.9	74.7	112	112	84	50.4	42
$\frac{(f_o - f_e)^2}{f_e}$	.338	.007	.080	2.580	1.190	17.384	11.524

$\chi^2 = 33.103 \qquad \chi^2_{CRIT} = \chi^2_{.05;6} = 12.592$

reject $H_0 \Rightarrow$ Poisson distribution with $\lambda = 3$ is not appropriate

11-26 a)

	A	B	C	D	E
	4.4	5.8	4.8	2.9	4.6
	4.6	5.2	5.9	2.7	4.3
	4.5	4.9	4.9	2.9	3.8
	4.1	4.7	4.6	3.9	5.2
	3.8	4.6	4.3	4.3	4.4
$\sum x$	21.4	25.2	24.5	16.7	22.3

n	5	5	5	5	5
$\bar{x}$	4.28	5.04	4.90	3.34	4.46
$\sum x^2$	92.02	127.94	121.51	57.81	100.49
s^2	0.107	0.233	0.365	0.508	0.258

grand mean $= \bar{\bar{x}} = \dfrac{21.4 + 25.2 + 24.5 + 16.7 + 22.3}{25} = 4.404$

b) $\hat{\sigma}_b^2 = \dfrac{\sum n_i(\bar{x}_i - \bar{\bar{x}})^2}{k-1} = \dfrac{9.0056}{4} = 2.2514$

c) $\hat{\sigma}_w^2 = \dfrac{\sum (n_i - 1)s_i^2}{n_T - k} = \dfrac{5.884}{20} = 0.2942$

d) $F = \dfrac{2.2514}{0.2942} = 7.65$

$F_{.05}(4,20) = 2.87$, so reject $H_0 \Rightarrow$ the brands produce significantly different amounts of relief

11-28

	n	$\bar{x}$	s^2
Employee 1	4	14.5	4.333
Employee 2	4	13	8.667
Employee 3	5	13	2.5
Employee 4	6	11.667	3.467

grand mean $= \bar{\bar{x}} = \dfrac{4(14.5) + 4(13) + 5(13) + 6(11.667)}{19} = 12.8947$

$\hat{\sigma}_b^2 = \dfrac{19.4562}{3} = 6.4854$

$\hat{\sigma}_w^2 = \dfrac{66.3333}{15} = 4.4222$

$F = \dfrac{6.4854}{4.4222} = 1.47$

$F_{.05}(3,15) = 3.29$, so do not reject $H_0 \Rightarrow$ the employees' mean performances are not significantly different

Here : $H_0 : \mu_1 = \mu_2 = \mu_3 = \mu_4$
H_1 : at least one pair $\mu_i - \mu_j \neq 0$

11-30 a)

	n	$\bar{x}$	s^2
Speed 1	5	36	2.5
Speed 2	5	31	7
Speed 3	5	35	10
Speed 4	5	31	10

$\bar{\bar{x}} = \dfrac{5(36) + 5(31) + 5(35) + 5(31)}{5 + 5 + 5 + 5} = 33.25$

b) $\hat{\sigma}_b^2 = \dfrac{\sum n_i(\bar{x}_i - \bar{\bar{x}})^2}{k-1} = \dfrac{103.75}{3} = 34.5833$

c) $\hat{\sigma}_w^2 = \dfrac{\sum (n_i - 1)s_i^2}{n_T - k} = \dfrac{118}{16} = 7.375$

d) $F = \dfrac{\hat{\sigma}_b^2}{\hat{\sigma}_w^2} = \dfrac{34.5833}{7.375} = 4.69$

$F_{CRIT} = F_{.05}(3,16) = 3.24$, so reject $H_0 \Rightarrow$ the different speeds do significantly affect the number of defective clocks

11-32

	n	$\bar{x}$	s^2
November	6	46.6667	81.0667
December	6	46.1667	47.7667
January	6	33.5	21.5

$$\bar{\bar{x}} = \frac{6(46.667) + 6(46.1667) + 6(33.5)}{6 + 6 + 6} = 42.1111$$

$$\hat{\sigma}_b^2 = \frac{\sum n_i(\bar{x}_i - \bar{\bar{x}})^2}{k-1} = \frac{668.1146}{2} = 334.0573$$

$$\hat{\sigma}_w^2 = \frac{\sum (n_i - 1)s_i^2}{n_T - k} = \frac{751.6667}{15} = 50.1111$$

$$F = \frac{\hat{\sigma}_b^2}{\hat{\sigma}_w^2} = \frac{334.0573}{50.1111} = 6.67$$

$F_{CRIT} = F_{.05}(2,15) = 3.68$, so reject H_0 $\Rightarrow$ the number of shoplifters differs significantly from month to month

11-34

	n	$\bar{x}$	s^2
Room 1	5	5.7	1.45
Room 2	5	4.1	1.55
Room 3	5	2.4	1.425
Room 4	5	7.6	1.675
Room 5	5	2.2	1.075

$$\bar{\bar{x}} = \frac{5(5.7) + 5(4.1) + 5(2.4) + 5(7.6) + 5(2.2)}{5 + 5 + 5 + 5 + 5} = 4.4$$

$$\hat{\sigma}_b^2 = \frac{\sum n_i(\bar{x}_i - \bar{\bar{x}})^2}{k-1} = \frac{104.3}{4} = 26.075$$

$$\hat{\sigma}_w^2 = \frac{\sum (n_i - 1)s_i^2}{n_T - k} = \frac{28.7}{20} = 1.435$$

$$F = \frac{\hat{\sigma}_b^2}{\hat{\sigma}_w^2} = \frac{26.075}{1.435} = 18.17$$

$F_{CRIT} = F_{.05}(4,20) = 2.87$, so reject H_0 $\Rightarrow$ at least two clean rooms have significantly different numbers of dust particles

11-36

	n	$\bar{x}$	s^2
Generic	5	18.0	38.0000
DNA	4	19.5	99.6667
RNA	6	16.5	23.9000
Oops	5	19.4	41.3000

$$\bar{\bar{x}} = \frac{5(18.0) + 4(19.5) + 6(16.5) + 5(19.4)}{5 + 4 + 6 + 5} = 18.2$$

$$\hat{\sigma}_b^2 = \frac{\sum n_i(\bar{x}_i - \bar{\bar{x}})^2}{k-1} = \frac{31.5}{3} = 10.5$$

$$\hat{\sigma}_w^2 = \frac{\sum (n_i - 1)s_i^2}{n_T - k} = \frac{735.7}{16} = 45.9813$$

$$F = \frac{\hat{\sigma}_b^2}{\hat{\sigma}_w^2} = \frac{10.5}{45.9813} = 0.23$$

$F_{CRIT} = F_{.05}(3,16) = 3.24$, so do not reject H_0 $\Rightarrow$ the quantities of jeans sold don't differ across brands

11-38

	n	$\overline{x}$	s^2
Restaurant 1	5	4	0.875
Restaurant 2	5	4.1	0.925
Restaurant 3	5	4.6	3.425
Restaurant 4	5	3.6	1.425

a) $\overline{\overline{x}} = \dfrac{5(4) + 5(4.1) + 5(4.6) + 5(3.6)}{5 + 5 + 5 + 5} = 4.075$

$$\hat{\sigma}_b^2 = \frac{\sum n_i(\overline{x}_i - \overline{\overline{x}})^2}{k-1} = \frac{2.5375}{3} = 0.8458$$

$$\hat{\sigma}_w^2 = \frac{\sum (n_i - 1)s_i^2}{n_T - k} = \frac{26.6}{16} = 1.6625$$

$$F = \frac{\hat{\sigma}_b^2}{\hat{\sigma}_w^2} = \frac{0.8458}{1.6625} = 0.51$$

$F_{CRIT} = F_{.05}(3.16) = 3.24$, so do not reject H_0 $\Rightarrow$ the restaurants do not have significantly different service times

b) Since no group stands out as having a mean greater than any of the others, the owner has no reason to single out any of the manager's restaurants for improvement needed more than the other restaurants.

11-40 $H_0: \sigma = 50$ $\quad$ $H_1: \sigma \neq 50$

The observed $\chi^2 = \dfrac{(n-1)s^2}{\sigma^2} = \dfrac{29(57)^2}{50^2} = 37.688$

$\chi^2_{.975;29} = 16.047$ $\quad$ $\chi^2_{.025;29} = 45.722$

Do not reject H_0 $\Rightarrow$ sample standard deviation of 57 is not significantly different from the hypothesized standard deviation of 50

11-42 a) $H_0: \sigma = 2$ (or $\sigma^2 = 4$) $\quad$ $H_1: \sigma < 2$ (or $\sigma^2 < 4$)

b) observed $\chi^2 = \dfrac{(n-1)s^2}{\sigma^2} = \dfrac{29(1.46)^2}{4} = 15.4541$

$\chi^2_{CRIT} = \chi^2_{.99;29} = 14.256$, so do not reject H_0

c) The observed s^2 of 1.46 is not significantly below 2, so the telescope should not be sold.

11-44 a) $H_0: \sigma^2 = 64$
$H_1: \sigma^2 \neq 64$

b) $\chi^2 = \dfrac{(n-1)s^2}{\sigma^2} = \dfrac{19(28)}{64} = 8.31$

$\chi^2_{.975;19} = 8.907$ $\quad$ $\chi^2_{.025;19} = 32.852$ $\quad$ Reject H_0

c) Thus, six-year olds' attention span is significantly different in variability from five-year olds' attention span.

11-46 $H_0: \sigma^2 = 80$ (no change)
$H_1: \sigma^2 < 80$ (the change has reduced variance)

observed $\chi^2 = \dfrac{(n-1)s^2}{\sigma^2} = \dfrac{24(28)}{80} = 8.4$

$\chi^2_{CRIT} = \chi^2_{.95;24} = 13.848$

Thus, reject H_0 $\Rightarrow$ the new policy does reduce the variance significantly

11-48 $n_1 = 16$ $s_1^2 = 3.75$ $n_2 = 10$ $s_2^2 = 5.38$

$H_0 : \sigma_1^2 = \sigma_2^2$

$H_1 : \sigma_1^2 \neq \sigma_2^2$

Test at $\alpha = .10$, with 15,9 df

a) $F = \frac{s_1^2}{s_2^2} = \frac{3.75}{5.38} = 0.70$

b) $F_{CRIT} = F_{.05}(15,9) = 3.01$

c) $F_{.95}(15,9) = \frac{1}{F_{.05}(9,15)} = \frac{1}{2.59} = 0.39$

d) do not reject H_0

11-50 $n_1 = 25$ $s_1 = 15.0$ $n_2 = 14$ $s_2 = 9.7$

$H_0 : \sigma_1 = \sigma_2$ (or $\sigma_1^2 = \sigma_2^2$)
$H_1 : \sigma_1 > \sigma_2$ (or $\sigma_1^2 > \sigma_2^2$)

observed $F = \frac{s_1^2}{s_2^2} = \frac{15.0^2}{9.7^2} = 2.39$

$F_{CRIT} = F_{.01}(24,13) = 3.59$

Thus, do not reject H_0 $\Rightarrow$ we conclude that the variance of the second sample is not significantly smaller

11-52 $n_A = 20$ $s_A = 0.6$ $n_B = 25$ $s_B = 0.9$

$H_0 : \sigma_B^2 = \sigma_A^2$

$H_1 : \sigma_B^2 > \sigma_A^2$

observed $F = \frac{s_B^2}{s_A^2} = \frac{0.81}{0.36} = 2.25$

$F_{CRIT} = F_{.01}(24,19) = 2.92$

Thus, we do not reject H_0 $\Rightarrow$ patients at hospital A do not have significantly less variability in their recovery times.

11-54 $H_0 : \sigma_{PAL}^2 = \sigma_{CAL}^2$ $H_1 : \sigma_{PAL}^2 > \sigma_{CAL}^2$

observed $F = \frac{s_{PAL}^2}{s_{CAL}^2} = \frac{20^2}{10^2} = 4$

$F_{CRIT} = F_{.05}(24,24) = 1.98$

Thus, reject $H_0 \Rightarrow$ PAL's processing speed is significantly more variable than CAL's

11-56 $H_0 = \sigma_Y^2 = \sigma_G^2$ $H_1 = \sigma_Y^2 \neq \sigma_G^2$

observed $F = \frac{s_Y^2}{s_G^2} = \frac{16}{10} = 1.6$

$F_{.05}(24,10) = 2.74$ $F_{.95}(24,10) = \frac{1}{F_{.05}(10,24)} = \frac{1}{2.25} = 0.44$

Thus, do not reject H_0 $\Rightarrow$ no significant difference in the variance of ice cream weights between Yum-Yum and Goody

11-58 H_0: occupation and attitude toward social legislation are independent
H_1: occupation and attitude toward social legislation are dependent

f_o	f_e	$f_o - f_e$	$(f_o - f_e)^2$	$\frac{(f_o - f_e)^2}{f_e}$
19	18.8108	0.1892	0.0358	0.0019
16	15.8919	0.1081	0.0117	0.0007
37	37.2973	−0.2973	0.0884	0.0024
15	21.6847	−6.6847	44.6852	2.0607
22	18.3198	3.6802	13.5439	0.7393
46	42.9955	3.0045	9.0270	0.2100
24	17.5045	6.4955	42.1915	2.4103
11	14.7883	−3.7883	14.3512	0.9704
32	34.7072	−2.7072	7.3289	0.2112

$$\chi^2 = \sum \frac{(f_o - f_e)^2}{f_e} = 6.6069$$

$\chi^2_{CRIT} = \chi^2_{.05;4} = 9.488$

Do not reject H_0 $\Rightarrow$ occupation and attitudes toward social legislation appear to be unrelated

11-60 a) normal b) χ^2

c) analysis of variance (F-distribution) d) t-test

11-62 H_0 : the proportion of patents granted has not changed over the 10 year period
H_1 : the proportion of patents granted has changed over the 10 year period

f_o	f_e	$f_o - f_e$	$(f_o - f_e)^2$	$\frac{(f_o - f_e)^2}{f_e}$
39223	36641.0241	2581.9759	6666599.548	169.9666
51183	53767.9494	− 2584.5374	6681963.400	130.5504
26548	29129.9759	− 2581.9759	6666599.548	251.1149
45331	42746.0506	2584.9494	6681963.400	147.4038

$$\chi^2 = \sum \frac{(f_o - f_e)^2}{f_e} = 699.0357$$

$\chi^2_{CRIT} = \chi^2_{.05;1} = 3.841$,

so we reject H_0 $\Rightarrow$ There was a significant change in the proportion of patents awarded over the last 10 years.

11-64 a) t-test (t-distribution) b) F-distribution

c) normal d) χ^2

11-66

	n	$\bar{x}$	s^2
Billboard 1	8	34.0000	94.571
Billboard 2	9	28.5556	64.528
Billboard 3	7	32.7143	100.238

$$\bar{\bar{x}} = \frac{8(34) + 9(28.5556) + 7(32.7143)}{8 + 9 + 7} = 31.5833$$

$$\hat{\sigma}_b^2 = \frac{\sum n_i(\bar{x}_i - \bar{\bar{x}})^2}{k-1} = \frac{138.1803}{2} = 69.09 \qquad \hat{\sigma}_w^2 = \frac{\sum (n_i - 1)s_i^2}{n_T - k} = \frac{1779.649}{21} = 84.745$$

$$F = \frac{\hat{\sigma}_b^2}{\hat{\sigma}_w^2} = \frac{69.09}{84.745} = 0.82$$

$F_{CRIT} = F_{.05}(2,21) = 3.47$, so do not reject H_0 $\Rightarrow$ the three traffic volumes are not significantly different

11-68 a, b)

f_o	f_e	$f_o - f_e$	$(f_o - f_e)^2$	$\frac{(f_o - f_e)^2}{f_e}$
27	22.2	4.8	23.04	1.0378
48	55.5	−7.5	56.25	1.0135
15	12.3	2.7	7.29	0.5927
25	25.16	−0.16	0.0256	0.0010
63	62.9	0.1	0.01	0.0002
14	13.94	0.06	0.0036	0.0003
22	26.64	−4.64	21.5296	0.8082
74	66.6	7.4	54.76	0.8222
12	14.76	−2.76	7.6176	0.5161

$$\chi^2 = \sum \frac{(f_o - f_e)^2}{f_e} = 4.7920$$

c) H_0: church attendance and income level are independent
H_1: church attendance and income level are dependent

d) $\chi^2_{CRIT} = \chi^2_{.05;4} = 9.488$, so accept H_0 $\Rightarrow$ church attendance seems to be unrelated to income level

11-70

	n	$\overline{x}$	s^2
Industrial	30	0.2417	0.3932
Transportation	20	0.0813	0.3343
Utility	15	0.2000	0.0844

$$\overline{\overline{x}} = \frac{30(.2417) + 20(.0813) + 15(.2)}{30 + 20 + 15} = 0.1827$$

$$\hat{\sigma}_b^2 = \frac{\sum n_i(\overline{x}_i - \overline{\overline{x}})^2}{k-1} = \frac{.3145}{2} = 0.1573$$

$$\hat{\sigma}_w^2 = \frac{\sum (n_i - 1)s_i^2}{n_T - k} = \frac{19.748}{62} = 0.3185$$

$$F = \frac{\hat{\sigma}_b^2}{\hat{\sigma}_w^2} = \frac{.1573}{.3185} = .4939$$

$F_{CRIT} = F_{.05}\,(2,62) = 3.15$,

so do not reject H_0 $\Rightarrow$ the three groups did not have significantly different average changes in share prices on that particular day.

11-72 a) Let $x =$ the number of correct guesses
H_0 : x is distributed binomially ($n = 10$, $p = .5$)
H_1 : x is not distributed as above

b)

# of shifts	0-2	3-5	6-10
Binomial prob.	.0547	.5683	.3770
Observed	50	47	3
Expected	5.47	56.83	37.70

$\dfrac{(f_o - f_e)^2}{f_e}$	362.5084	1.7003	31.9387

(The last two categories are combined because $f_e = 1.07$ for 9-10 correct guesses.)

$$\chi^2_{OBS} = \sum \frac{(f_0 - f_e)^2}{f_e} = 396.147 \qquad \chi^2_{CRIT} = \chi^2_{.10;2} = 4.605$$

reject H_0 $\Rightarrow$ Swami Zhami's probability of guessing the correct card is not .5

c)

# of correct guesses	0-2	3-10
Binomial probability	.5256	.4744
Observed	50	50
Expected	52.56	47.44
$\dfrac{(f_o - f_e)^2}{f_e}$	.1247	.1381

(The last three categories are combined because $f_e < 5$.)

$$\chi^2_{OBS} = \sum \frac{(f_0 - f_e)^2}{f_e} = .2628 \qquad \chi^2_{CRIT} = \chi^2_{.10;1} = 2.706$$

do not reject H_0 $\Rightarrow$ Swami Zhami has no psychic power ($p = .25$)

11-74 H_0 : Jim's errors are $N(0,16)$

H_1 : Jim's errors follow another distribution

Errors	< -6.5	-6.5 to 0.5	0.5 to 6.5	> 6.51
Z	-1.63	0.13	1.63	
Normal prob.	.0516	.5001	.3967	.0516
Observed	5	45	45	5
Expected	5.16	50.01	39.67	5.16
$\dfrac{(f_o - f_e)^2}{f_e}$	.0050	.5019	.7161	.0050

$$\chi^2 = \sum \frac{(f_o - f_e)^2}{f_e} = 1.228 \qquad \chi^2_{CRIT} = \chi^2_{.05;3} = 7.815$$

do not reject H_0 $\Rightarrow$ Jim's errors are normally distributed with mean 0 and variance 16

11-76

	n	$\bar{x}$	s^2
NC	11	32.8727	4.1422
SA	11	34.8000	28.3660
SC	9	30.3889	3.7911

$$\bar{\bar{x}} = \frac{11(32.8727) + 11(34.8) + 9(30.3889)}{11 + 11+9} = 32.8355$$

$$\hat{\sigma}^2_b = \frac{\sum n_i(\bar{x}_i - \bar{\bar{x}})^2}{k-1} = \frac{96.3398}{2} = 48.1699$$

$$\hat{\sigma}^2_w = \frac{\sum (n_i - 1)s_i^2}{n_T - k} = \frac{359.2019}{28} = 12.8286$$

$$F = \frac{\hat{\sigma}^2_b}{\hat{\sigma}^2_w} = \frac{48.1699}{12.8286} = 3.7549$$

$F_{CRIT} = F_{.05}\,(2,28) = 3.35$,

so reject H_0 $\Rightarrow$ the average ages in these regions do differ significantly.

11-78 H_0: Normal distribution H_1: Some other distribution

No μ or σ are specified, so we'll estimate them by $\bar{x} = 1,764,857.8$ and $s = 409,322.2$. As suggested in exercise 11-17, we'll use 5 equally probable intervals, so $df = 5-1-2 = 2$. No significance level is specified.

The 20th, 40th, 60th, and 80th percentiles of any distribution divide that distribution into five equally probable intervals. For the standard normal distribution, those percentiles are $z = -0.84, -0.25, 0.25$, and 0.84. Since $x = \mu + \sigma z$, the corresponding percentiles for a normal distribution with $\mu = 1,764,857.8$ and $\sigma = 409,322.2$ are:

20th percentile: $1,764,857.8 - 409,322.2(0.84) = 1,421,027.2 = A$
40th percentile: $1,764,857.8 - 409,322.2(0.25) = 1,662,527.3 = B$
60th percentile: $1,764,857.8 + 409,322.2(0.25) = 1,867,188.4 = C$
80th percentile: $1,764,857.8 + 409,322.2(0.84) = 2,108,688.4 = D$

Sales interval	$< A$	A to B	B to C	C to D	$> D$
Observed	9	11	12	10	8
Expected	10	10	10	10	10
$\frac{(f_o - f_e)^2}{f_e}$	0.1	0.1	0.4	0.0	0.4

$\chi^2 = 1.0$; even with α as large as 0.20 (with $\chi^2_{CRIT} = \chi^2_{.20;2} = 3.219$) we would not reject H_0. Hence, the retail-sales data are well described by a normal distribution.

11-80

	n	$\bar{x}$	s^2
Drug 1	4	245.75	72.9167
Drug 2	4	272.50	41.6667
Drug 3	5	229.00	138.5000
Drug 4	5	243.40	39.3000

$$\bar{\bar{x}} = \frac{4.(245.75) + 4(272.5) + 5(229) + 5(243.4)}{4+4+5+5} = 246.3889$$

$$\hat{\sigma}_b^2 = \frac{\sum n_i(\bar{x}_i - \bar{\bar{x}})^2}{k-1} = \frac{4285.3278}{3} = 1428.4426$$

$$\hat{\sigma}_w^2 = \frac{\sum (n_i - 1)s_i^2}{n_T - k} = \frac{1054.95}{14} = 75.3536$$

$$F = \frac{\hat{\sigma}_b^2}{\hat{\sigma}_w^2} = \frac{1428.4426}{75.3536} = 18.96$$

$F_{CRIT} = F_{.05}(3,14) = 3.34$, so reject $H_0 \Rightarrow$ the four drugs do affect driving skill differently

11-82

	n	$\bar{x}$	s^2
Aircraft Type A	5	7.6	0.345
Aircraft Type B	3	6.8	1.120
Aircraft Type C	6	8.7	0.404

$$\bar{\bar{x}} = \frac{5(7.6) + 3(6.8) + 6(8.7)}{5+3+6} = 7.9$$

$$\hat{\sigma}_b^2 = \frac{\sum n_i(\bar{x}_i - \bar{\bar{x}})^2}{k-1} = \frac{7.92}{2} = 3.96$$

$$\hat{\sigma}_w^2 = \frac{\sum(n_i-1)s_i^2}{n_T-k} = \frac{5.64}{11} = 0.513$$

$$F = \frac{\hat{\sigma}_b^2}{\hat{\sigma}_w^2} = \frac{3.96}{0.513} = 7.72$$

$F_{CRIT} = F_{.01}(2,11) = 7.21$, so reject H_0 $\Rightarrow$ there are significantly different fuel costs per mile among the three types of aircraft

11-84 H_0: $\sigma_A^2 = \sigma_N^2$ (variability the same in both leagues)

H_0: $\sigma_A^2 > \sigma_N^2$ (more variability in the American League)

$n_A = 26$ $\sigma_A^2 = 0.001499$ $n_N = 24$ $\sigma_N^2 = 0.001372$

$F = \frac{\sigma_A^2}{\sigma_N^2} = 1.09$, with a prob-value of 0.4171 for this upper-tailed test.

Since this is greater than our significance level of $\alpha = 0.10$, we do not reject H_0. Batting skills are not significantly more variable in the American League. Dick is correct.

CHAPTER 12

SIMPLE REGRESSION AND CORRELATION

NOTE: In hand-calculated regressions, all intermediate results are carried to only 4 decimal places, resulting in rounding errors in some of the solutions. For some problems, 4-place solutions from SAS are also given for comparison purposes.

12-2 An estimating equation is the formula describing the relationship between a dependent variable and one or more independent variables.

12-4 In a "direct (indirect)" relationship the dependent variable increases (decreases) as the independent variable increases.

12-6 In a linear relationship, the dependent variable changes a constant amount for equal incremental changes in the independent variable(s). In a curvi-linear relationship, the dependent variable does not change at a constant rate with equal incremental changes in the independent variable(s).

12-8 It is a process which determines the relationship between a dependent variable and more than one independent variable.

12-10 a) We want to see if Final Average (FA) depends on Quiz Average (QA), so FA is the dependent variable and QA is the independent variable.

b)

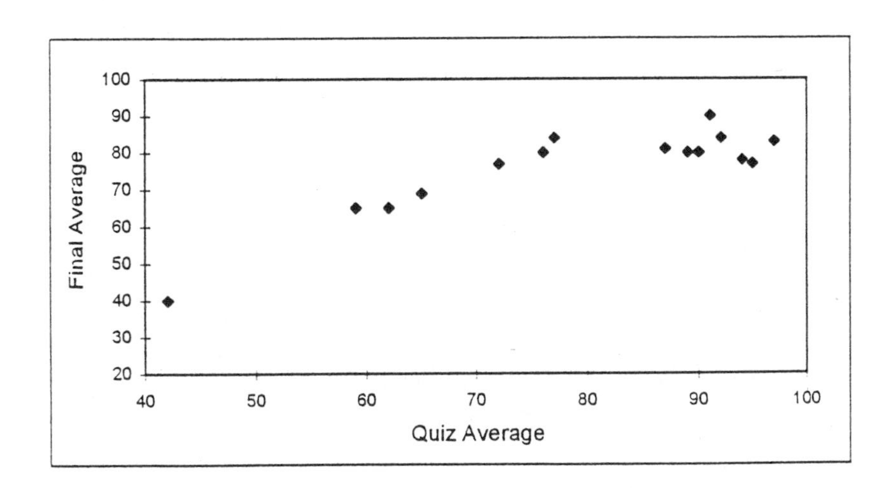

c) curvilinear

d) For the most part, the professor is right. However, for very high quiz averages, the curve appears to be turning down.

12-12

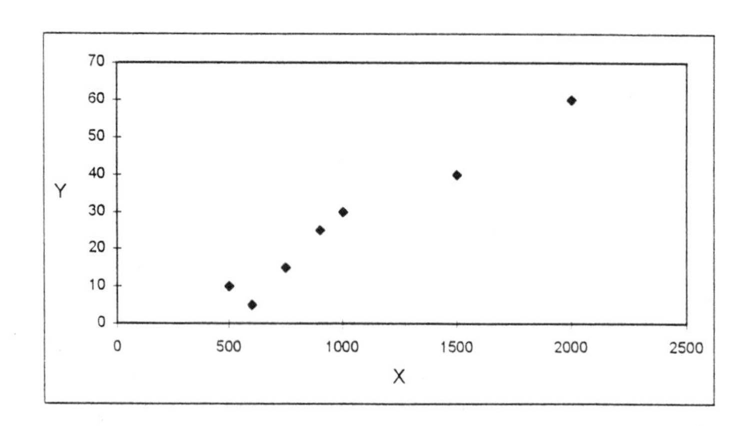

A direct linear relationship seems to exist. Clearly it is absurd to suggest that the use of facial tissues causes the common cold. (But the opposite may well be true.)

12-14 a)

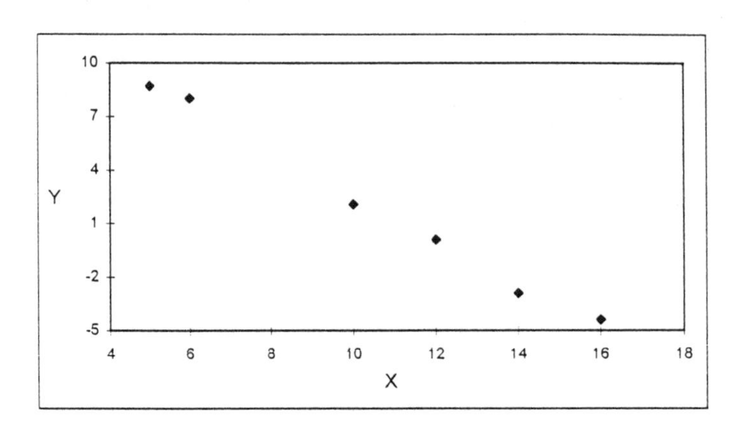

b)

X	Y	XY	X^2
16	−4.4	−70.4	256
6	8.0	48.0	36
10	2.1	21.0	100
5	8.7	43.5	25
12	0.1	1.2	144
14	−2.9	−40.6	196
$\sum X = 63$	$\sum Y = 11.6$	$\sum XY = 2.7$	$\sum X^2 = 757$

$\overline{X} = \frac{63}{6} = 10.5$ $\qquad$ $\overline{Y} = \frac{11.6}{6} = 1.9333$

$$b = \frac{\sum XY - n\overline{XY}}{\sum X^2 - n\overline{X}^2} = \frac{2.7 - 6(10.5)(1.9333)}{757 - 6(10.5)^2} = -1.2471$$

$a = \overline{Y} - b\overline{X} = 1.9333 - (-1.2471)(10.5) = 15.0279$

Thus, $\widehat{Y} = 15.0279 - 1.2471X$ $\qquad$ (SAS: $\widehat{Y} = 15.0281 - 1.2471X$)

c) $X = 5,\ \widehat{Y} = 15.0279 - 1.2471(5) = 8.7924$

$X = 6,\ \widehat{Y} = 15.0279 - 1.2471(6) = 7.5453$

$X = 7,\ \widehat{Y} = 15.0279 - 1.2471(7) = 6.2982$

12-16 a)

X(housing starts)	Y(appliance sales)	XY	X^2	Y^2
2.0	5.0	10.00	4.00	25.00
2.5	5.5	13.75	6.25	30.25
3.2	6.0	19.20	10.24	36.00
3.6	7.0	25.20	12.96	49.00
3.3	7.2	23.76	10.89	51.84
4.0	7.7	30.80	16.00	59.29
4.2	8.4	35.28	17.64	70.56
4.6	9.0	41.40	21.16	81.00
4.8	9.7	46.56	23.04	94.09
5.0	10.0	50.00	25.00	100.00
$\sum X = 37.2$	$\sum Y = 75.5$	$\sum XY = 295.95$	$\sum x^2 = 147.18$	$\sum Y^2 = 597.03$

$\overline{X} = \frac{37.2}{10} = 3.72$ $\qquad$ $\overline{Y} = \frac{75.5}{10} = 7.55$

$$b = \frac{\sum XY - n\overline{XY}}{\sum X^2 - n\overline{X}^2} = \frac{295.95 - 10(3.72)(7.55)}{147.18 - 10(3.72)^2} = 1.7156$$

$a = \overline{Y} - b\overline{X} = 7.55 - 1.7156(3.72) = 1.1680$

Thus, $\widehat{Y} = 1.1680 + 1.7156X$ (SAS: $\widehat{Y} = 1.1681 + 1.7156X$)

b) When housing starts go up 1000 units, appliance sales go up 1.7156 thousand units.

c) $$s_e = \sqrt{\frac{\sum Y^2 - a\sum Y - b\sum XY}{n-2}} = \sqrt{\frac{597.03 - 1.1680(75.5) - 1.7156(295.95)}{8}}$$

$= 0.3732$ (SAS: 0.3737)

(The standard deviation at the data points around the regression line is about 370 appliances.)

d) $\widehat{Y} = 1.1680 + 1.7156(8) = 14.89$

$\widehat{Y} \pm t_{\alpha/2;n\text{-}2}\, s_e = 14.89 \pm 1.860(.3732) = 14.89 \pm .694 = (14.20, 15.58)$ thousand units.

12-18 In this problem, Y = passengers per 100 miles and X = ticket price.

a)

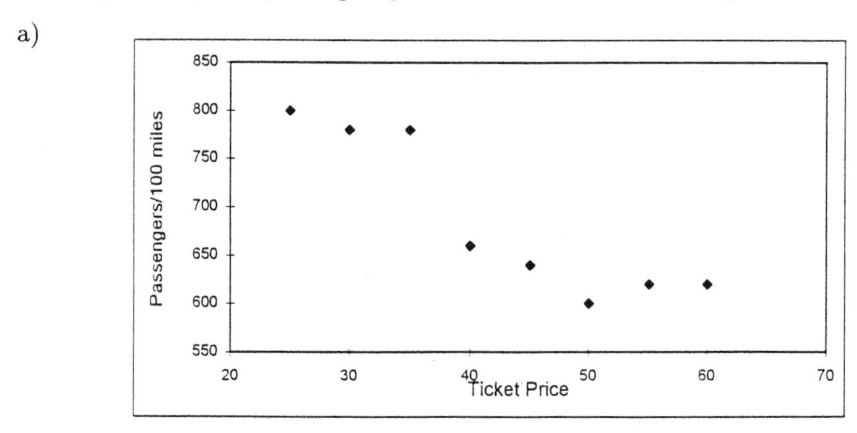

b)

X	Y	XY	X^2	Y^2
25	800	20000	625	640000
30	780	23400	900	608400
35	780	27300	1225	608400
40	660	26400	1600	435600
45	640	28800	2025	409600

50	600	30000	2500	360000
55	620	34100	3025	384400
60	620	37200	3600	384400
$\sum X = 340$	$\sum Y = 5500$	$\sum XY = 227200$	$\sum X^2 = 15500$	$\sum Y^2 = 3830800$

$\overline{X} = 340/8 = 42.5$ $\qquad\qquad$ $\overline{Y} = 5500/8 = 687.5$

$$b = \frac{\sum XY - n\overline{X}\overline{Y}}{\sum X^2 - n\overline{X}^2} = \frac{227200 - 8(42.5)(687.5)}{15500 - 8(42.5)^2} = -6.2381$$

$a = \overline{Y} - b\overline{X} = 687.5 - (-6.2381)(42.5) = 952.6193$

Thus, $\widehat{Y} = 952.6193 - 6.2381X$ (SAS: $\widehat{Y} = 952.6190 - 6.2381X$)

c)

$$s_e = \sqrt{\frac{\sum Y^2 - a\sum Y - b\sum XY}{n-2}}$$

$$= \sqrt{\frac{3830800 - 952.6193(5500) - (-6.2381)(227200)}{6}} = 38.0573 \text{ (SAS: 38.0580)}$$

$\widehat{Y} = 952.6193 - 6.2381(50) = 640.7143$

$\widehat{Y} \pm t_{\alpha/2;n\text{-}2}s_e = 640.7143 \pm 2.447(38.0573)$

$= 640.7143 \pm 93.1262 = (547.5881, 733.8405)$

12-20 In this problem, Y = degree of arousal and X = noise level.

a)

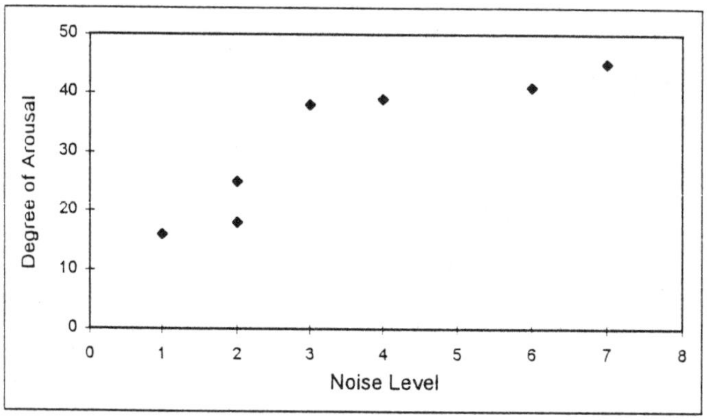

b)

X	Y	XY	X^2
4	39	156	16
3	38	114	9
1	16	16	1
2	18	36	4
6	41	246	36
7	45	315	49
2	25	50	4
3	38	114	9
$\sum X = 28$	$\sum Y = 260$	$\sum XY = 1047$	$\sum X^2 = 128$

$\overline{X} = \frac{28}{8} = 3.5$ $\qquad\qquad$ $\overline{Y} = \frac{260}{8} = 32.5$

$$b = \frac{\sum XY - n\overline{X}\overline{Y}}{\sum X^2 - n\overline{X}^2} = \frac{1047 - 8(3.5)(32.5)}{128 - 8(3.5)^2} = 4.5667$$

$a = \overline{Y} - b\overline{X} = 32.5 - 4.5667(3.5) = 16.5166$

Thus, $\widehat{Y} = 16.5166 + 4.5667X$ (SAS: $\widehat{Y} = 16.5167 - 4.5667X$)

c) $\widehat{Y} = 16.5166 + 4.5667(5) = 39.35$

12-22 In this problem, Y = number of accidents per game, X = number of games played.

a)

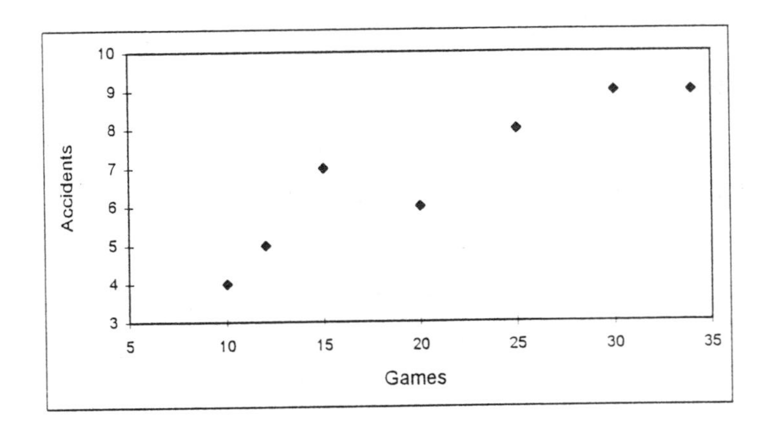

b)

X	Y	XY	X^2	Y^2
20	6	120	400	36
30	9	270	900	81
10	4	40	100	16
12	5	60	144	25
15	7	105	225	49
25	8	200	625	64
34	9	306	1156	81
$\sum X = 146$	$\sum Y = 48$	$\sum XY = 1101$	$\sum X^2 = 3550$	$\sum Y^2 = 352$

$\overline{X} = \frac{146}{7} = 20.8571$ $\overline{Y} = \frac{48}{7} = 6.8571$

$$b = \frac{\sum XY - n\overline{X}\overline{Y}}{\sum X^2 - n\overline{X}^2} = \frac{1101 - 7(20.8571)(6.8671)}{3550 - 7(20.8571)^2} = 0.1978$$

$a = \overline{Y} - b\overline{X} = 6.8571 - 0.1978(20.8571) = 2.7316$

Thus, $\widehat{Y} = 2.7316 + 0.1978X$ (SAS: $\widehat{Y} = 2.7317 + 0.1978X$)

c) $\widehat{Y} = 2.7316 + 0.1978(33) = 9.3$

d) $$s_e = \sqrt{\frac{\sum Y^2 - a\sum Y - b\sum XY}{n-2}}$$

$$= \sqrt{\frac{352 - 2.7317(48) - 0.1978(1101)}{5}} = 0.7875 \quad \text{(SAS: 0.7882)}$$

12-24 In this problem, Y = percent of pollutants and X = money spent.

a)

X	Y	XY	X^2	Y^2
8.4	35.9	301.56	70.56	1288.81
10.2	31.8	324.36	104.04	1011.24
16.5	24.7	407.55	272.25	610.09
21.7	25.2	546.84	470.89	635.04
9.4	36.8	345.92	88.36	1354.24
8.3	35.8	297.14	68.89	1281.64
11.5	33.4	384.10	132.25	1115.56
18.4	25.4	467.36	338.56	645.16
16.7	31.4	524.38	278.89	985.96
19.3	27.4	528.82	372.49	750.76
28.4	15.8	448.72	806.56	249.64
4.7	31.5	148.05	22.09	992.25
12.3	28.9	355.47	151.29	835.21
$\sum X = 185.8$	$\sum Y = 384.0$	$\sum XY = 5080.27$	$\sum X^2 = 3177.12$	$\sum Y^2 = 11755.60$

$\overline{X} = \frac{185.8}{13} = 14.2923$ $\qquad \overline{Y} = \frac{384.0}{13} = 29.5385$

$$b = \frac{\sum XY - n\overline{XY}}{\sum X^2 - n\overline{X}^2} = \frac{5080.27 - 13(14.2923)(29.5385)}{3177.12 - 13(14.2923)^2} = -0.7822$$

$a = \overline{Y} - b\overline{X} = 29.5385 + 0.7822(14.2923) = 40.7179$

Thus, $\widehat{Y} = 40.7179 - 0.7822X$ (SAS: $\widehat{Y} = 40.7172 - 0.7821X$)

b) $\widehat{Y} = 40.7179 - 0.7822(20) = 25.0739$, so 25.0739 percent of the emissions will be dangerous pollutants.

c) $$s_e = \sqrt{\frac{\sum Y^2 - a\sum Y - b\sum XY}{n-2}} = \sqrt{\frac{11755.6 - 40.7179(384.0) - (-0.7822)(5080.27)}{11}}$$
$= 2.9188$

12-26 $$r^2 = \frac{a\sum Y + b\sum XY - n\overline{Y}^2}{\sum Y^2 - n\overline{Y}^2} = \frac{34.6470(25) + (-5.2941)(131) - 6(5)^2}{175 - 5(5)^2} = 0.9530 \text{ (SAS:0.9529)}$$

$r = -\sqrt{0.9530} = -0.9762$

12-28

Y^2
3364
1681
2025
729
676
144
256
9
8884

$$r^2 = \frac{a\sum Y + b\sum XY - n\overline{Y}^2}{\sum Y^2 - n\overline{Y}^2}$$

$$= \frac{70.5(228) + (-2.8)(2580) - 8(28.5)^2}{8884 - 8(28.5)^2} = .9858$$

$r = -\sqrt{.9858} = -.9929$

12-30 $$r^2 = \frac{a\sum Y + b\sum XY - n\overline{Y}^2}{\sum Y^2 - n\overline{Y}^2}$$

$$= \frac{-4.6151(1371) + 41.6809(4954) - 10(137.1)^2}{201121 - 10(137.1)^2} = 0.9269 \quad \text{(SAS: .9270)}$$

$r = \sqrt{0.9269} = 0.9628$

12-32 a)

X	Y	XY	X^2	Y^2
3	11	33	9	121
7	18	126	49	324
4	9	36	16	81
2	4	8	4	16
0	7	0	0	49
4	6	24	16	36
1	3	3	1	9
2	8	16	4	64
$\sum X = 23$	$\sum Y = 66$	$\sum XY = 246$	$\sum X^2 = 99$	$\sum Y^2 = 700$

$\overline{X} = 23/8 = 2.875$ $\qquad \overline{Y} = 66/8 = 8.25$

$$b = \frac{\sum XY - n\overline{XY}}{\sum X^2 - n\overline{X}^2} = \frac{246 - 8(2.875)(8.25)}{99 - 8(2.875)^2} = 1.7110$$

$a = \overline{Y} - b\overline{X} = 8.25 - 1.7110(2.875) = 3.3309$

Thus, $\widehat{Y} = 3.3309 + 1.7110X$ *(SAS:* $\widehat{Y} = 3.3308 + 1.7110X$)

b) $r^2 = \dfrac{a\sum Y + b\sum XY - n\overline{Y}^2}{\sum Y^2 - n\overline{Y}^2} = \dfrac{3.3309(66) + 1.7110(246) - 8(8.25)^2}{700 - 8(8.25)^2} = 0.6189$

$r = \sqrt{0.6189} = 0.7867$

12-34

X	Y	XY	X^2	Y^2
3.6	12.13	43.668	12.96	147.1369
4.8	14.70	70.560	23.04	216.0900
9.7	22.83	221.451	94.09	521.2089
12.6	28.40	357.840	158.76	806.5600
11.5	28.33	325.795	132.25	802.5889
10.9	27.05	294.845	118.81	731.7025
14.6	33.60	490.560	213.16	1128.9600
18.2	40.80	742.560	331.24	1664.6400
3.7	9.40	34.780	13.69	88.3600
9.8	24.84	243.432	96.04	617.0256
12.4	30.17	374.108	153.76	910.2289
16.9	34.70	586.430	285.61	1204.0900
$\sum X = 128.7$	$\sum Y = 306.95$	$\sum XY = 3786.029$	$\sum X^2 = 1633.41$	$\sum Y^2 = 8838.5920$

$\overline{X} = \dfrac{128.7}{12} = 10.7250$ $\qquad \overline{Y} = \dfrac{306.95}{12} = 25.5792$

$$b = \frac{\sum XY - n\overline{X}\overline{Y}}{\sum X^2 - n\overline{X}^2} = \frac{3786.029 - 12(10.7250)(25.5792)}{1633.41 - 12(10.7250)^2} = 1.9517$$

$a = \overline{Y} - b\overline{X} = 25.5792 - 1.9517(10.7250) = 4.6472$ (SAS: $a = 4.6468$, $b = 1.9517$)

$$s_e = \sqrt{\frac{\sum Y^2 - a\sum Y - b\sum XY}{n-2}}$$

$$= \sqrt{\frac{8838.592 - 4.6472(306.95) - 1.9517(3786.029)}{10}} = 1.5146 \quad \text{(SAS: 1.5141)}$$

$$s_b = \frac{s_e}{\sqrt{\sum X^2 - n\overline{X}^2}} = \frac{1.5146}{\sqrt{1633.41 - 12(10.7250)^2}} = 0.0952$$

$H_0: B = 1.5$ $\qquad H_1: B > 1.5$ $\qquad \alpha = .05$

The upper limit of the acceptance region is $B + t(s_b) = 1.5 + 1.812(0.0952) = 1.67$
Here, $b = 1.9517 > 1.67 \Rightarrow$ reject $H_0 \Rightarrow$ Ned should advertise.

12-36 $H_0: B = 1.50$ $\qquad H_1: B \neq 1.50$ $\qquad \alpha = .05$

The limits of the acceptance region are $B - t(s_b)$ $= 1.5 \pm 2.069(.11)$
$= 1.5 \pm .23 = [1.27, 1.73]$

Here, $b = 1.685 \Rightarrow$ do not reject $H_0 \Rightarrow$ slope has not changed significantly from its past value.

12-38 a) The 90% confidence interval is $b \pm t(s_b)$ $= .147 \pm 1.746(.032)$
$= .147 \pm .0559 = [.091, .203]$

Since .08 is not contained within this confidence interval, it does appear, at the .10 significance level, that the true slope has changed from the value found in 1969.

b) The 99% confidence interval is $b \pm t(s_b)$ $= .147 \pm 2.921(.032)$
$= .147 \pm .0935 = [.054, .241]$

Since .08 is contained within this confidence interval, it does not appear, at the .01 significance level, that the true slope has changed from the value found in 1969.

12-40 $H_0: B = 0.85$ $\quad H_1: B \neq 0.85$ $\quad \alpha = .01$

$$s_b = \frac{s_e}{\sqrt{\sum X^2 - n\overline{X}^2}} = \frac{0.60}{\sqrt{0.25}} = 1.2$$

$B \pm t(s_b) = .85 \pm 2.878(1.2) = .85 \pm 3.45 = [-2.6, 4.3]$
Here, $b = 0.70 \Rightarrow$ do not reject $H_0 \Rightarrow$ the slope has not changed significantly.

12-42 The coefficient of determination is the fraction of the variation in the dependent variable which is explained by the independent variable. The coefficient of correlation only shows whether the relationship is direct or inverse; it does not tell how much of the variation in Y is explained by X.

12-44 Correlation does not imply causality because it is simply a statistical technique applies to a set of numbers thought to show a relationship between variables. The relationship between two variables may be due to an external cause affecting them both. Only by the manipulation of variables and then observation of the results can we infer any kind of causality.

12-46 a) Let Y = storks and X = babies

X	Y	XY	X^2	Y^2
27	35	945	729	1225
38	46	1748	1444	2116
13	19	247	169	361
24	32	768	576	1024
6	15	90	36	225
19	31	589	361	961
15	20	300	225	400
$\sum X = 142$	$\sum Y = 198$	$\sum XY = 4687$	$\sum X^2 = 3540$	$\sum Y^2 = 6312$

$\overline{X} = \frac{142}{7} = 20.2857$ $\qquad \overline{Y} = \frac{198}{7} = 28.2857$

$$b = \frac{\sum XY - n\overline{X}\overline{Y}}{\sum X^2 - n\overline{X}^2} = \frac{4687 - 7(20.2857)(28.2857)}{3540 - 7(20.2857)^2} = 1.0167$$

$a = \overline{Y} - b\overline{X} = 28.2857 - 1.0167(20.2857) = 7.6612$ (SAS: 7.6616)

$$r^2 = \frac{a\sum Y + b\sum XY - n\overline{Y}^2}{\sum Y^2 - n\overline{Y}^2}$$

$$= \frac{7.6612(198) + 1.0167(4687) - 7(28.2857)^2}{6312 - 7(28.2857)^2} = 0.9581$$

$r = \sqrt{.9581} = 0.9788$

b) No. The high correlation is spurious. It simply reflects the fact that the number of storks and the number of babies tend to rise together when the population increases. Higher population means more births and more nesting sites for storks.

12-48
$$r^2 = \frac{a\sum Y + b\sum XY - n\overline{Y}^2}{\sum Y^2 - n\overline{Y}^2} = \frac{15.0279(11.6) + (-1.2471)(2.7) - 6(1.9333)^2}{171.88 - 6(1.9333)^2} = 0.9938$$

$r = -\sqrt{0.9938} = -0.9969$

12-50

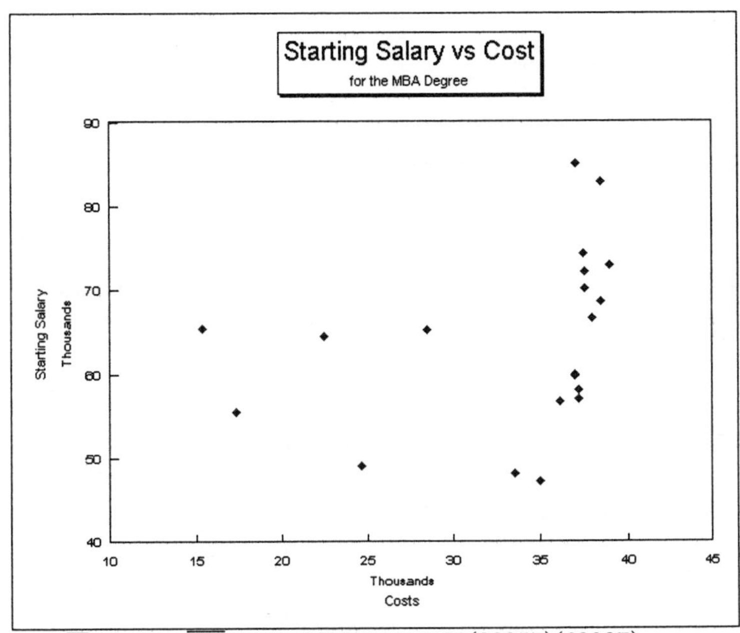

$$b = \frac{\sum XY - n\overline{X}\overline{Y}}{\sum X^2 - n\overline{X}^2} = \frac{43024221800 - 20(33257)(63987)}{23156470000 - 20(33257)^2} = 0.4478$$

$$a = \overline{Y} - b\overline{X} = 63987 - .4478(33257) = 49094.5154$$

$$\hat{Y} = 49094.5154 + 0.4478X$$

$$s_e = \sqrt{\frac{\sum Y^2 - a\sum Y - b\sum XY}{n-2}}$$

$$s_e = \sqrt{\frac{83999286400 - 49094.5154(1279740) - .4478(43024221800)}{18}} = 10287.0327$$

$$s_b = \frac{s_e}{\sqrt{\sum X^2 - n\overline{X}^2}} = \frac{10287.0327}{\sqrt{23156470000 - 20(33257)^2}} = 0.3196$$

$H_0 : B = 0$ $\quad$ $H_1 : B > 0$ $\quad$ $\alpha = .05$

$$t = \frac{b - B_{H_0}}{s_b} = \frac{-0.4478 - 0}{0.3196} = 1.4011$$

$$t_{CRIT} = t_{.05, 18} = 2.101$$

Since $t = 1.4011$ (which is less than 2.101), do not reject H_0. Greater school costs are not associated with higher salaries.

12-52 Business Week overall rankings (From 12-51):

$b = -1508.5113$
$a = 75101.3687$

$$r^2 = \frac{a\sum Y + b\sum XY - n\overline{Y}^2}{\sum Y^2 - n\overline{Y}^2}$$

$$r^2 = \frac{75101.37(1279740) + (-1508.51)(12733360) - 20(63987)^2}{83999276400 - 20(63987)^2} = 0.3527$$

Student rankings:

$$b = \frac{\sum XY - n\overline{X}\overline{Y}}{\sum X^2 - n\overline{X}^2} = \frac{13595200 - 20(11)(63987)}{3286 - 20(11)^2} = -556.513$$

$$a = \overline{Y} - b\overline{X} = 63987 - (-556.513)(11) = 70108.64$$

$$r^2 = \frac{a\sum Y + b\sum XY - n\overline{Y}^2}{\sum Y^2 - n\overline{Y}^2}$$

$$r^2 = \frac{70108.64(1279740) + (-556.513)(13595200) - 20(63987)^2}{83999276400 - 20(63987)^2} = 0.1267$$

Firm rankings:

$$b = \frac{\sum XY - n\overline{X}\overline{Y}}{\sum X^2 - n\overline{X}^2} = \frac{12807650 - 20(10.5)(63987)}{2870 - 20(10.5)^2} = -946.797$$

$$a = \overline{Y} - b\overline{X} = 63987 - (-949.797)(10.5) = 73928.37$$

$$r^2 = \frac{a\sum Y + b\sum XY - n\overline{Y}^2}{\sum Y^2 - n\overline{Y}^2}$$

$$r^2 = \frac{73928.37(1279740) + (-946.797)(12807650) - 20(63987)^2}{83999276400 - 20(63987)^2} = 0.2822$$

The overall ranking explains the largest fraction of the variation in starting salaries.

12-54 a) 1, + b) 2, + c) 2, − d) 2, −

12-56 $$r^2 = \frac{a\sum Y + b\sum XY - n\overline{Y}^2}{\sum Y^2 - n\overline{Y}^2} = \frac{4.9205(809.48) + 3.9650(12688.48) - 15(53.9653)^2}{51156.28 - 15(53.9653)^2} = 0.9613$$

$$r = \sqrt{0.9613} = 0.9805$$

12-58 SALES and POP

$$b = \frac{\sum XY - n\overline{X}\overline{Y}}{\sum X^2 - n\overline{X}^2} = \frac{20994383720 - 50(1764858)(235.324)}{2810359 - 50(235.324)^2} = 5512.5812$$

$$a = \overline{Y} - b\overline{X} = 1764858 - 5512.5812(235.324) = 467615.1589$$

$$r^2 = \frac{a\sum Y + b\sum XY - n\overline{Y}^2}{\sum Y^2 - n\overline{Y}^2}$$

$$r^2 = \frac{467615.1589(88242891) + 5512.5812(2099438720) - 50(1764858)^2}{163945843262647 - 50(1764858)^2} = 0.1536$$

SALES and TM

$$b = \frac{\sum XY - n\overline{X}\overline{Y}}{\sum X^2 - n\overline{X}^2} = \frac{650886990558 - 50(7180.379)(1764858)}{2704284838 - 50(7180.379)^2} = 136.6344$$

$$a = \overline{Y} - b\overline{X} = 1764858 - 136.6344(7180.379) = 783771.369$$

$$r^2 = \frac{a\sum Y + b\sum XY - n\overline{Y}^2}{\sum Y^2 - n\overline{Y}^2}$$

$$r^2 = \frac{783771.369(88242891) + 136.6344(650886990558) - 50(1764858)^2}{163945843262647 - 50(1764858)^2} = 0.2874$$

TM explains more of the variation in SALES than does POP.

12-60 SALES and AGE

$\overline{X} = 32.774 \qquad \overline{Y} = 1764857.82$

$$b = \frac{\sum XY - n\overline{X}\overline{Y}}{\sum X^2 - n\overline{X}^2} = \frac{2900658083 - 50(32.774)(1764857.82)}{54317.81 - 50(32.774)^2} = 14049.8998$$

$$a = \overline{Y} - b\overline{X} = 1764857.82 - 14049.8998(32.774) = 1304386.404$$

$$\hat{Y} = 1304386.404 + 14049.9X$$

$$s_e = \sqrt{\frac{\sum Y^2 - a\sum Y - b\sum XY}{n-2}}$$

$$s_e = \sqrt{\frac{163945843262647 - 1304386.404(88242891) - 14049.8998(2900658083)}{50-2}} = 410514.332$$

$$s_b = \frac{s_e}{\sqrt{\sum X^2 - n\overline{X}^2}} = \frac{410514.322}{\sqrt{54317.81 - 50(32.774)^2}} = 16606.87$$

$H_0 : B = 0 \qquad H_1 : B > 0$

$$t = \frac{b - B_{H_0}}{s_b} = \frac{14049.9 - 0}{16606.87} = 0.846$$

With 48 df, the probability value for the test is greater than 10%, so we would probably accept H_0. Although this appears to indicate that "Business isn't better in communities with lots of older people", it would be erroneous to draw such a conclusion. As we saw in Exercise 12-58, POP explains 15% of the variation in SALES, and a simple regression of SALES on AGE ignores this factor. In order to legitimately draw the suggested conclusion, you should first do a multiple regression analysis.

12-62 a

12-64 In this problem, Y = rent and X = number of bedrooms.

a)

X	Y	XY	X^2	Y^2
2	230	460	4	52900
1	190	190	1	36100
3	450	1350	9	202500
2	310	620	4	96100
2	218	436	4	47524
2	185	370	4	34225
2	340	680	4	115600
1	245	245	1	60025
1	125	125	1	15625
2	350	700	4	122500
2	280	560	4	78400
$\sum X = 20$	$\sum Y = 2923$	$\sum XY = 5736$	$\sum X^2 = 40$	$\sum Y^2 = 861499$

$\overline{X} = 20/11 = 1.81818 \qquad \overline{Y} = 2923/11 = 265.72727$

$$b = \frac{\sum XY - n\overline{X}\overline{Y}}{\sum X^2 - n\overline{X}^2} = \frac{5736 - 11(1.81818)(265.72727)}{40 - 11(1.81818)^2} = 115.8991$$

$$a = \overline{Y} - b\overline{X} = 265.72727 - 115.8991(1.81818) = 55.0018$$

Thus, $\hat{Y} = 55.0018 + 115.8991X$ (SAS: $\hat{Y} = 55.0 + 115.9X$)

b) $$r^2 = \frac{a\sum Y + b\sum XY - n\overline{Y}^2}{\sum Y^2 - n\overline{Y}^2} = \frac{55.0018(2923) + 115.8991(5736) - 11(265.72727)^2}{861499 - 11(265.72727)^2}$$

$$= 0.5762$$

c) $\widehat{Y} = 55.0018 + 115.8991(2) = 286.80$

12-66 $b = \dfrac{\sum XY - n\overline{X}\overline{Y}}{\sum X^2 - n\overline{X}^2} = \dfrac{74.3961 - 10(2.571)(2.946)}{68.6733 - 10(2.571)^2} = -0.523$

$a = \overline{Y} - b\overline{X} = 2.946 - (-0.523)(2.571) = 4.2906$

$\widehat{Y} = 4.2906 - 0.523X$

Answer: (a) Lower price increases sales

12-68 In this problem, Y = price and X = size of offering.

a)

X	Y	XY	X^2	Y^2
108.0	12.0	1296.00	11664.00	144.00
4.4	4.0	17.60	19.36	16.00
3.5	5.0	17.50	12.25	25.00
3.6	6.0	21.60	12.96	36.00
39.0	13.0	507.00	1521.00	169.00
68.4	19.0	1299.60	4678.56	361.00
7.5	8.5	63.75	56.25	72.25
5.5	5.0	27.50	30.25	25.00
375.0	15.0	5625.00	140625.00	225.00
12.0	6.0	72.00	144.00	36.00
51.0	12.0	612.00	2601.00	144.00
66.0	12.0	792.00	4356.00	144.00
10.4	6.5	67.60	108.16	42.25
4.0	3.0	12.00	16.00	9.00
$\sum X = 758.3$	$\sum Y = 127$	$\sum XY = 10431.15$	$\sum X^2 = 165844.79$	$\sum Y^2 = 1448.50$

$\overline{X} = 758.3/14 = 54.16429$ $\qquad \overline{Y} = 127/14 = 9.07143$

$b = \dfrac{\sum XY - n\overline{X}\overline{Y}}{\sum X^2 - n\overline{X}^2} = \dfrac{10431.15 - 14(54.16429)(9.07143)}{165844.79 - 14(54.16429)^2} = 0.0285$

$a = \overline{Y} - b\overline{X} = 9.07143 - 0.0285(54.16429) = 7.5277$

Thus, $\widehat{Y} = 7.5277 + 0.0285X$ (SAS: $\widehat{Y} = 7.5294 + 0.0285X$)

b) $r^2 = \dfrac{a\sum Y + b\sum XY - n\overline{Y}^2}{\sum Y^2 - n\overline{Y}^2} = \dfrac{7.5277(127) + 0.0285(10431.15) - 14(9.07143)^2}{1448.5 - 14(9.07143)^2}$

$= 0.3415$ (SAS: 0.3412)

Since only about a third of the variability in price is explained by the size of the offering, Dave should search for additional explanatory variables.

12-70 In this problem, Y = adult height (in inches) and X = birth weight (in ounces).

X	Y	XY	X^2	Y^2
88	69	6072	7744	4761
112	72	8064	12544	5184
100	66	6600	10000	4356
120	71	8520	14400	5041
130	73	9490	16900	5329
108	70	7560	11664	4900
$\sum X = 658$	$\sum Y = 421$	$\sum XY = 46306$	$\sum X^2 = 73252$	$\sum Y^2 = 29571$

$\overline{X} = 658/6 = 109.6667$ $\qquad \overline{Y} = 421/6 = 70.1667$

$$b = \frac{\sum XY - n\overline{X}\overline{Y}}{\sum X^2 - n\overline{X}^2} = \frac{46306 - 6(109.6667)(70.1667)}{73252 - 6(109.6667)^2} = 0.1249$$

$$a = \overline{Y} - b\overline{X} = 70.1667 - 0.1249(109.6667) = 56.4693$$

Thus, $\widehat{Y} = 56.4693 + 0.1249X$ (SAS: $\widehat{Y} = 56.4667 + 0.1249X$)

$$r^2 = \frac{a\sum Y + b\sum XY - n\overline{Y}^2}{\sum Y^2 - n\overline{Y}^2} = \frac{56.4693(421) + 0.1249(46306) - 6(70.1667)^2}{29571 - 6(70.1667)^2}$$

$$= 0.5519 \quad \text{(SAS: 0.5524)}$$

Thus, 55% of variation in adult height is explained by birth weight.

12-72 a) Using a calculator, we find $\sum X = 317.271$, $\sum Y = 910.612$, $\sum XY = 29,221.8450$, $\sum X^2 = 10,246.2891$, and $\sum Y^2 = 85,344.9104$. Hence

$\overline{X} = 317.271/10 = 31.7271$ $\overline{Y} = 910.612/10 = 91.0612$

$$b = \frac{\sum XY - n\overline{X}\overline{Y}}{\sum X^2 - n\overline{X}^2} = \frac{29,221.8450 - 10(31.7271)(91.0612)}{10,246.2891 - 10(31.7271)^2} = 1.8356$$

$$a = \overline{Y} - b\overline{X} = 91.0612 - 1.8356(31.7172) = 32.8229$$

Thus, $\widehat{Y} = 32.8229 + 1.8356X$ (SAS: $\widehat{Y} = 32.8245 + 1.8356X$)

$$s_e = \sqrt{\frac{\sum Y^2 - a\sum Y - b\sum XY}{n - 2}}$$

$$= \sqrt{\frac{85,344.9104 - 32.8229(910.612) - 1.8356(29,221.8450)}{10 - 2}} = 15.0680$$

$$s_b = \frac{s_e}{\sqrt{\sum X^2 - n\overline{X}^2}} = \frac{15.0680}{\sqrt{10,246.2891 - 10(31.7271)^2}} = 1.1225$$

$H_0: B = 0$ $H_1: B \neq 0$ $\alpha = 0.05$

The limits of the acceptance region are $B \pm ts_b = 0 \pm 2.306(1.1225) = \pm 2.5885$. Here, $b = 1.8356 < 2.5885$, so do not reject H_0. The Attorney General's salary is not related to the going rate for attorneys in the state.

b) $$r^2 = \frac{a\sum Y + b\sum XY - n\overline{Y}^2}{\sum Y^2 - n\overline{Y}^2} = \frac{32.8229(910.612) + 1.8356(29,221.8450) - 10(91.0612)^2}{85,344.9104 - 10(91.0612)^2}$$

$$= 0.2505$$

Thus, 25% of the variation in AG's salaries is accounted for by the going rate for attorneys in the for-profit market.

c) No. First of all, as we've just seen, the relationship between the two is not particularly strong. But even if r^2 had been higher, remember that r^2 measures association, not causality.

CHAPTER 13

MULTIPLE REGRESSION AND MODELING TECHNIQUES

13-2 To enable us to include qualitative factors as explanatory variables in regression models.

13-4 Yes. Qualitative factors such as season of the year can be modeled using the techniques of "dummy variables."

13-6 No. Conceptually they are quite similar since multiple regression is based on the same assumptions as simple regression. However, they will be more complex in terms of the computations that need to be done.

13-8 a)

Y	X_1	X_2	X_1Y	X_2Y	X_1X_2	X_1^2	X_2^2	Y^2
10	8	4	80	40	32	64	16	100
17	21	9	357	153	189	441	81	289
18	14	11	252	198	154	196	121	324
26	17	20	442	520	340	289	400	676
35	36	13	1260	455	468	1296	169	1225
8	9	28	72	224	252	81	784	64
114	105	85	2463	1590	1435	2367	1571	2678

Equations 11-2, 3, and 4 become

$$\sum Y = na + b_1 \sum X_1 + b_2 \sum X_2 \qquad 114 = 6a + 105b_1 + 85b_2$$

$$\sum X_1 Y = a \sum X_1 + b_1 \sum X_1^2 + b_2 \sum X_1 X_2 \qquad 2463 = 105a + 2367b_1 + 1435b_2$$

$$\sum X_2 Y = a \sum X_2 + b_1 \sum X_1 X_2 + b_2 \sum X_2^2 \qquad 1590 = 85a + 1435b_1 + 1571b_2$$

Solving these equations simultaneously, we get

$a = 2.5915$ $\quad b_1 = 0.8897$ $\quad b_2 = 0.0592$

So the regression equation is

$$\hat{Y} = 2.5915 + 0.8897X_1 + 0.0592X_2$$

b) With $X_1 = 28$ and $X_2 = 10$,

$$\hat{Y} = 2.5915 + 0.8897(28) + 0.0592(10) = 28.10$$

13-10 a) In this problem, $Y =$ sales, $X_1 =$ advertising, $X_2 =$ price.

Y	X_1	X_2	X_1Y	X_2Y	X_1X_2	X_1^2	X_2^2	Y^2
33	3	125	99	4125	375	9	15625	1089
61	6	115	366	7015	690	36	13225	3721
70	10	140	700	9800	1400	100	19600	4900
82	13	130	1066	10660	1690	169	16900	6724
17	9	145	153	2465	1305	81	21025	289
24	6	140	144	3360	840	36	19600	576
287	47	795	2528	37425	6300	431	105975	17299

Equations 13-2, 3, and 4 become

$\sum Y = na + b_1 \sum X_1 + b_2 \sum X_2$	$287 = 6a + 47b_1 + 795b_2$
$\sum X_1 Y = a \sum X_1 + b_1 \sum X_1^2 + b_2 \sum X_1 X_2$	$2528 = 47a + 431b_1 + 6300b_2$
$\sum X_2 Y = a \sum X_2 + b_1 \sum X_1 X_2 + b_2 \sum X_2^2$	$37425 = 795a + 6300b_1 + 105975b_2$

Solving these equations simultaneously, we get

$a = 219.2306 \qquad b_1 = 6.3815 \qquad b_2 = -1.6708$

So the regression equation is

$$\hat{Y} = 219.2306 + 6.3815X_1 - 1.6708X_2$$

b) When advertising = 7 and price = 132,

$$\hat{Y} = 219.2306 + 6.3815(7) - 1.6708(132) = 43.33 \text{ units}$$

13-12 In this problem, Y = price, X_1 = year, X_2 = miles.

Y	X_1	X_2	X_1Y	X_2Y	X_1X_2	X_1^2	X_2^2	Y^2
2.99	1987	55.6	5941.13	166.244	110477.2	3948169	3091.36	8.9401
6.02	1992	18.4	11991.84	110.768	36652.8	3968064	338.56	36.2404
8.87	1993	21.3	17677.91	188.931	42450.9	3972049	453.69	78.6769
3.92	1988	46.9	7792.96	183.848	93237.2	3952144	2199.61	15.3664
9.55	1994	11.8	19042.70	112.690	23529.2	3976036	139.24	91.2025
9.05	1991	36.4	18018.55	329.420	72472.4	3964081	1324.96	81.9025
9.37	1992	28.2	18665.04	264.234	56174.4	3968064	795.24	87.7969
4.20	1988	44.2	8349.60	185.640	87869.6	3952144	1953.64	17.6400
4.80	1989	34.9	9547.20	167.520	69416.1	3956121	1218.01	23.0400
5.74	1991	26.4	11428.34	151.536	52562.4	3964081	696.96	32.9476
64.51	19905	324.1	128455.30	1860.831	644842.2	39620953	12211.27	473.7533

Equations 13-2, 3, and 4 become

$\sum Y = na + b_1 \sum X_1 + b_2 \sum X_2$	$64.51 = 10a + 19905b_1 + 324.1b_2$
$\sum X_1 Y = a \sum X_1 + b_1 \sum X_1^2 + b_2 \sum X_1 X_2$	$128455.3 = 19905a + 39620953b_1 + 644842.2b_2$
$\sum X_2 Y = a \sum X_2 + b_1 \sum X_1 X_2 + b_2 \sum X_2^2$	$1860.831 = 324.1a + 644842.2b_1 + 12211.27b_2$

Solving these equations simultaneously, we get

$a = -4243.1682 \qquad b_1 = 2.1315 \qquad b_2 = 0.2135$

So the regression equation is

$$\hat{Y} = -4243.1682 + 2.1315X_1 + 0.2135X_2$$

b) When year = 1991 and miles = 40.0,

$$\hat{Y} = -4243.1682 + 2.1315(1991) + 0.2135(40) = \$9.188 \text{ (in thousands)}$$

13-14 Results taken from computer output:

a) $\hat{Y} = 34.8079 + 5.2618X_1 - 8.0187X_2 + 6.8084X_3$

b) $s_e = 4.0688$

c) $R^2 = .9834$

d) $\hat{Y} = 34.8079 + 5.2618(5.8) - 8.0187(4.2) + 6.8084(5.1) = 66.37$

13-16 Results taken from computer output:

a) $\widehat{Y} = 142.4363 + 3.2741X_1 + 0.5269X_2 - 0.3203X_3$

b) $R^2 = 0.9854 \Rightarrow$ 98.54% of total variation in Y explained by model

c) $\widehat{Y} = 142.4363 + 3.2741(82) + 0.5269(75.0) - 0.3203(10.5) = 453.8$

13-18 Results taken from computer output:

a) Predicted GRADE = −49.95 + 1.07HOURS + 1.36IQ + 2.04BOOKS − 1.80AGE

b) $R^2 = (.857)^2 = .7672$ (i.e., 76.72%)

c) Predicted GRADE = −49.95 + 1.07(5) + 1.36(113) + 2.04(3) − 1.80(21) = 77.40
The grade should be about 77.

13-20 Results taken from computer output:

a) Predicted SALESPRICE = −1.381 + 2.852SQFT − 3.713STORIES + 30.285BATHROOMS + 1.172AGE

b) $R^2 = 0.952 \Rightarrow$ 95.2% of the variation in SALESPRICE is explained by this regression.

c) When SQFT = 18, STORIES = 1, BATHROOMS = 1.5, and AGE = 6,
Predicted SALESPRICE = −1.381 + 2.852(18) − 3.713(1) + 30.285(1.5) + 1.172(6) = 98.7, i.e., $98,700

13-22 a) $H_0 : B_{DISPLAY} = 3 \qquad H_1 : B_{DISPLAY} < 3 \qquad \alpha = .05$

The lower limit of the acceptance region is:

$B_{DISPLAY} - t(S_{b_{DISPLAY}}) = 3 - 1.714(0.844) = 1.485$

Here, $b_{DISPLAY} = 1.25 < 1.485 \Rightarrow$ reject $H_0 \Rightarrow$ each display ad uses up significantly less than 3 pounds of newsprint (holding all else constant). Mark is wrong.

b) $H_0 : B_{CLASSIFIED} = 0.5 \qquad H_0 : B_{CLASSIFIED} \neq 0.5 \qquad \alpha = .05$

The limits of the acceptance region are:

$B_{CLASSIFIED} \pm t(s_{b_{CLASSIFIED}}) = 0.5 \pm 2.069(0.126) = 0.5 \pm 0.26 = [0.24, 0.76]$

Here, $b_{CLASSIFIED} = 0.251$ is within the acceptance region $\Rightarrow$ do not reject H_0
$\Rightarrow$ Mark is correct, each classified ad does use about 0.5 pounds of newsprint.

c) $H_0 : B_{FULLPAGE} = 333.333 \qquad H_1 : B_{FULLPAGE} > 333.333$

Here, $b_{FULLPAGE} = 250.659$ is clearly not significantly above 333.333

$\Rightarrow$ do not reject H_0 (without doing any formal test) $\Rightarrow$ Mark's rates are okay.

13-24 a) $F_{OBS} = \dfrac{\text{SSR}/k}{\text{SSE}/(n-k-1)} = \dfrac{783.604}{135.893} = 5.77$

b) $F_{CRIT} = F_{.05}(4,7) = 4.12$

c) $F_{OBS} > F_{CRIT} \Rightarrow$ reject H_0 that all B_i's = 0
$\Rightarrow$ Y depends on at least one of the X_i's
$\Rightarrow$ regression is significant as a whole.

13-26 H$_0$: $B_1 = B_2 = B_3 = B_4 = 0$ H$_1$: at least one $B_i \neq 0$ $\alpha = .05$

prob-value = .0001 < $\alpha = .05 \Rightarrow$ reject H$_0$ $\Rightarrow$ SALES depends on at least one of the independent variables
$\Rightarrow$ regression is significant as a whole

13-28 Juan would almost certainly run into the problem of multicollinearity, since the prime interest rate at banks is dependent upon the Federal Reserve's discount rate, which, for the most part, moves directly with the inflation rate. Thus, the estimates of the parameters in Juan's model would be very unreliable, and the model with both predictor variables included would probably not explain much more of the total variation in consumer demand than either of the simple regression models obtained using just one predictor variable.

13-30 Results taken from computer output:

a) Predicted TOURISTS = 5.9188 + 355470RATE − 0.1709PRICE + 0.2426PROMOT + 0.2273TEMP

b) H$_0$: $B_{RATE} = 0$ H$_1$: $B_{RATE} \neq 0$ $\alpha = .10$

prob-value = .2809 > $\alpha = .10 \Rightarrow$ do not reject H$_0$ $\Rightarrow$ the currency exchange rate is not a significant explanatory variable.

c) H$_0$: $B_{PROMOT} = 0.2$ H$_1$: $B_{PROMOT} > 0.2$ $\alpha = .05$

The upper limit of the acceptance region is:

$B_{PROMOT} \pm t(s_{b_{PROMOT}}) = 0.2 + 1.895(0.1628) = 0.5085$

Here, b_{PROMOT} = 0.2426 is within the acceptance region $\Rightarrow$ do not reject H$_0$
$\Rightarrow$ change in tourists for 1000 pound increase in promotions is not significantly more than 200.

d) A 95% confidence interval is:

$b_{TEMP} \pm t(s_{b_{TEMP}}) = 0.2273 \pm 2.365(0.1189)$
$= 0.2273 \pm 0.2812 = [-0.0539, 0.5085]$

13-32 a) Predicted REVENUE = $a + b_1$FLOW + b_2FLOW^2

b) Let CITY = 0 if restaurant is in one city and 1 for restaurant in the other city.
Predicted REVENUE = $a + b_1$FLOW + b_2FLOW^2 + b_3CITY

13-34 a) H$_0$: $B_1 = 0$ H$_1$: $B_1 \neq 0$ $\alpha = .05$

The limits of the acceptance region are:

$B_1 \pm t(s_{b_1}) = 0 \pm 2.110(3.245) = 0 \pm 6.85$

Here, $b_1 = 2.79$ is in the acceptance region $\Rightarrow$ do not reject H$_0$
$\Rightarrow$ X_1 is not a significant explanatory variable.

b) H$_0$: $B_2 = 0$ H$_1$: $B_2 \neq 0$ $\alpha = .05$

The limits of the acceptance region are:

$B_2 \pm t(s_{b_2}) = 0 \pm 2.110(1.53) = 0 \pm 3.23$

Here, $b_1 = -3.92 < -3.23 \Rightarrow$ reject H$_0$ $\Rightarrow$ X_2 $(= X_1^2)$ is a significant explanatory variable.

13-36 Results taken from computer output:

a) Predicted DEMAND = −0.9705 + 4.4146TIME

b) Predicted DEMAND = 3.4101 + 2.8686TIME + 0.0966TIME^2
This model is better than the linear model: R^2 has increased from 0.9886 to 0.9956, and both explanatory variables are highly significant (with prob-values of 0.0000 and 0.0009). Furthermore, the residuals in part (a) show a curvilinear pattern, but the residuals for this model are random.

13-38 a) The judge has spotted the obvious pattern in the residuals.

b) The most direct model is a second-degree equation. Add the square of the number of days in court as an additonal independent variable.

13-40 a) Predicted GROWTH = 70.066 + 0.422CREAT + 0.271MECH + 0.745ABST + 0.420MATH

b) $R^2 = 0.9261 \Rightarrow$ 92.61% of the variation in GROWTH is explained by this regression.

c)

Variable	Prob-value	Significant?
CREAT	0.0235	Yes
MECH	0.2284	No
ABST	0.0182	Yes
MATH	0.0001	Yes

Thus, the scores on the creativity, abstract thinking, and mathematical calculation aptitude tests are significant explanatory variables for sales growth.

d) PROB > $F = 0.0001$ very small $\Rightarrow$ yes, overall model is significant as a whole

e) Predicted GROWTH = 70.066 + 0.422(12) + 0.271(14) + 0.745(18) + 0.420(30) = 104.93

13-42 Results taken from computer output.

Predicted EATING = 56177.927 + 506.352POP
$R^2 = 0.0775$

Predicted EATING = 22170.308 + 5.029EBI
$R^2 = 0.258$

EBI accounts for more of the variation in EATING than POP. It explains 25.8% of the variation compared to 7.8% for POP.

13-44 Results taken from computer output.

Predicted EATING = − 104304.6 + 142.356POP − 4759.177SINGLE + 4.745EBI
$R^2 = 0.4419 = 44.19\%$ of the variation in EATING is explained by this model.

$b_{SINGLE} = 4759.177$

b_{SINGLE} is a significant explanatory variable ($t = 3.26$, p = .0021)

13-46 a) Predicted ANESTHES = 90.032 + 99.486TYPE + 21.536WEIGHT − 34.461HOURS

b) $\widehat{Y} = 90.032 + 99.486(1) + 21.536(25) - 34.461(1.5) = 676.23$ milliliters
An approximate 95% confidence interval is:

$$\widehat{Y} \pm t(s_e) = 676.23 \pm 2.262(57.070) = 676.23 \pm 129.09 = [547, 805]$$

c) $H_0 : B_{TYPE} = 0$ $\quad H_1 : B_{TYPE} \neq 0$ $\quad \alpha = .10$
prob-value = .0435 < $\alpha = .10$ $\Rightarrow$ reject H_0
$\Rightarrow$ there is a significant difference in the amount of anesthesia needed for dogs and cats (holding all else constant).

d) PROB $>$ $F = 0.0001$ very small $\Rightarrow$ yes, the regression is significant as a whole.

13-48 Results taken from computer output.

a) Predicted PRICE = 444.7183 − 0.6124WEIGHT − 4.3769SQFT

b) At WEIGHT = 100 ounces and SQFT = 46, the regression suggests a price of
444.7183 − 0.6124(100) − 4.3769(46) = $182.14

13-50 Results taken from computer output.

Predicted PRICE = − 5.7892 − 7.7134DIV + 3.8231EPS + 0.0352SALES
+ 0.0396INCOME − 0.018ASSETS + 1.5327OLDPR

$R^2 = 0.8043$ = 80.43% of the variation in PRICE is explained by this model.

13-52 Results taken from computer output.

Predicted PRICE = − 5.9374 − 9.9256DIV + 4.5839EPS + 1.4473OLDPR
+ 5.1679NY + 1.2772BANK

$R^2 = 0.8147$ = 81.47% of the variation in PRICE is explained by this model.

$H_0 : B_{NY} = 0$ $\qquad H_1 : B_{NY} \neq 0$

Since the prob-value = 0.137 > α = .10, we accept H_0. Being listed on the NYSE doesn't appear to have a significant effect on PRICE.

$H_0 : B_{BANK} = 0$ $\qquad H_1 : B_{BANK} \neq 0$

Since the prob-value = 0.772 > α = .10, we accept H_0. Neither share prices of banks nor bank holding companies differ significantly from those of other companies in the group.

13-54 Results taken from computer output.

a) Predicted REVENUE = 28725.416 − 139.760PROPERTY + 105.176SALES
+ 56.065GASOLINE

b) Predicted revenues under the two proposals are:

PROPOSAL A: 28725.416 − 139.760(2.763) + 105.176(1) + 56.065(1) = 28500.50

PROPOSAL B: 28725.416 − 139.760(1.639) + 105.176(2.021)) + 56.065(3.3) = 28893.50

If they wish to maximize their revenue, they should adopt proposal B, which yields $393,000 more than proposal A.

13-56 Results taken from computer output. The PHONES variable was recorded in 100,000's of units. The linear regression equation is:

Predicted PHONES = −6.6325 + 2.6040YEARS, with $r^2 = 0.7951$

The residuals (4.06, 1.77, −0.38, −2.35, −3.50, −3.91, −2.82, 0.51, and 6.62) distinctly show a pattern consistent with a quadratic curve. The quadratic regression equation is:

Predicted PHONES = 3.6280 − 2.9926YEARS + 0.5597(YEARS)^2,

with $r^2 = 0.9836$. This appears to be a much better fit to the data.

13-58 Results taken from computer output:

a) Predicted REVENUE = 8085.6084 + 51.4201STORES − 125.7441SIZE

Since the prob-value for STORES (0.0000) is smaller than that for SIZE (0.0132), the number of stores is more important in determining revenue growth. In fact, larger stores seem to be leading to a decline in revenues. This regression might lead a consultant to emphasize geographic spread.

b) With sales per employee measured in \$1000s,

$$\text{Predicted SALES/EMPLOYEE} = -40841.1511 + 20.8462\text{YEAR} - 5.1665\text{SIZE}$$

Because the coefficient of SIZE is negative, employees are not more productive in larger stores. The positive coefficient of YEARS shows that employee productivity is increasing over time. It appears that sales per employee is growing over time, notwithstanding the adverse effect of larger stores.

CHAPTER 14

NONPARAMETRIC METHODS

14-2 (b)

14-4 Nonparametric methods do not use all the information in the data, since they usually rely on ranks or counts, and they are not as efficient as are parametric tests in estimating and detecting things about parameters.

14-6 Yes, the company sacrifices a great deal of information by using a ranking test as its decision criterion. If the data were examined by graphing the number of preferences against the combination number, it could be seen that there is a very distinct bi-modal distribution. In this case, the choice of two benefit packages might well be the better alternative. If the company had chosen to use a very simple ranking test for ease of computation, it would have forfeited the information about the two distinct preferences of the participants in the sample.

14-8

Before	33	36	41	32	39	47	34	29	32	34	40	42	33	36	29
After	35	29	38	34	37	47	36	32	30	34	41	38	37	35	28
Sign	+	−	−	+	−	0	+	+	−	0	+	−	+	−	−

15 responses: 6(+); 7(−); 2(0)

For $n = 13$, $p = .5$, the expected number of +'s is 13(.5) = 6.5. We have observed 6 +'s. The probability of being this far or farther away from the expected value is $P(r \geq 6 \text{ or } r \geq 7) = 1$. Since $1 > .05$, we accept H_0. There has not been a significant change in collection time.

14-10 a) The meteorologist has a bit of a point, but it is not very strong. Even if 1995 is significantly cooler than 1994, that alone is not strong evidence of a long-run trend toward cooler weather.

b) 15 responses: 5 (+); 9(−); 1(0); (+ indicates a higher 1995 temperature)
For $n = 14$, $p = .5$, the probability of $\geq$ 9 −'s is .2120 (Appendix Table 3).
Since $.2120 > .05$, we must accept H_0. We conclude that there was not any significant cooling from 1994 to 1995.

14-12 For all parts, H_0 is $p = .5$ (same ideal sizes)
H_1 is $p > .5$ (mother's ideal greater)

(A + response indicates mother's ideal is greater than daughter's.)

a) 13 responses: 6(+); 4(−); 3(0)
For $n = 10$, $p = .5$, the probability of $\geq$ 6 +'s is .3770 (Appendix Table 3).
Since $.3770 \geq .03$, we would accept H_0.

b) $\sigma_{\bar{p}} = \sqrt{\frac{pq}{n}} = \sqrt{\frac{.5(.5)}{10}} = .1583$

At $\alpha = .03$, the upper limit of the acceptance region is:

$$.5 + 1.88\sigma_{\bar{p}} = .5 + 1.88(.1583) = .798$$

Since $\bar{p} = .6 < .798$, accept H_0. Daughters don't want significantly smaller families.

c) 143 responses: 66(+); 44(−); 33(0)

$$\sigma_{\overline{p}} = \sqrt{\frac{pq}{n}} = \sqrt{\frac{.5(.5)}{110}} = .0477$$

The upper limit of the acceptance region is:

$$.5 + 1.88(.0477) = .590$$

Since $\overline{p} = .6 > .590$, we reject H_0. The ideal family size has decreased.

d) Since $\sigma_{\overline{p}} = \sqrt{pq/n}$, the standard error gets smaller as n gets larger, and we put more trust in the data we have. It is easy to see that in this example, the percentages of smaller, larger, and no difference have not been altered. The only change in the computations of (b) and (c) is in the standard error of the mean. Hence when we took a sample of 143 and got 66 responses showing smaller we could conclude at $\alpha = .03$ that the ideal family size had decreased. With only 13 responses the reliability of our sample was lessened by its small size; hence we could not conclude at a significance level of $\alpha = .03$ that the ideal family size had decreased.

14-14

Men	31	25	38	33	42	40	44	26	43	35
Ranks	4	1	12	6	14	13	16.5	20	15	10
Women	44	30	34	47	35	32	35	47	48	34
Ranks	16.5	3	7.5	18.5	10	5	10	18.5	20	7.5

$n_1 = 10$ $\quad$ $n_2 = 10$ $\quad$ $\alpha = .10$

$R_1 = 93.5$ $\quad$ $R_2 = 116.5$

H_0: $\mu_1 = \mu_2$ $\quad$ H_1: $\mu_1 \neq \mu_2$

$$U = n_1 n_2 + \frac{n_1(n_1 + 1)}{2} - R_1 = 10(10) + \frac{10(11)}{2} - 93.5 = 61.5$$

$$\mu_U = \frac{n_1 n_2}{2} = \frac{10(10)}{2} = 50 \qquad \sigma_U = \sqrt{\frac{n_1 n_2(n_1 + n_2 + 1)}{12}} = \sqrt{\frac{10(10)(21)}{12}} = 13.23$$

The limits of the acceptance region are:

$$\mu_U \pm 1.96\sigma_U = 50 \pm 1.64(13.23) = 50 \pm 21.70 = (28.30,\ 71.70)$$

Thus, we accept H_0.

14-16

Credit cards	78	64	75	45	82	69	60
Ranks	17	8	16	1	18	12	6
Checks	110	70	53	51	61	68	
Ranks	20	13.5	3	2	7	10.5	
Cash	90	68	70	54	74	65	59
Ranks	19	10.5	13.5	4	15	9	5

$n_1 = 7$ $\quad$ $n_2 = 6$ $\quad$ $n_3 = 7$ $\quad$ $\alpha = .05$

$R_1 = 78$ $\quad$ $R_2 = 56$ $\quad$ $R_3 = 76$

H_0: $\mu_1 = \mu_2 = \mu_3$ $\quad$ H_1: the μ's are not all the same

$$K = \frac{12}{n(n + 1)} \sum \frac{R_j^2}{n_j} - 3(n + 1) = \frac{12}{20(21)}\left(\frac{78^2}{7} + \frac{56^2}{6} + \frac{76^2}{7}\right) - 3(21)$$

$$= 0.341$$

With $3 - 1 = 2$ degrees of freedom, the upper limit of the acceptance region is $\chi^2_{2,.05} = 5.991$, so we accept H_0. The average amounts paid by the three methods are not significantly different.

14-18

Old Machine	Ranks	New Machine	Ranks
992	21	965	17
945	15	1054	25
938	12.5	912	11
1027	24	850	2
892	6	796	1
983	20	911	10
1014	22	866	4
1258	26	902	9
966	18	956	16
889	5	900	8
972	19	938	12.5
940	14		
873	3		
1016	23		
897	7		

$n_1 = 15$ $\quad n_2 = 11$ $\quad \alpha = .10$

$R_1 = 235.5$ $\quad R_2 = 115.5$

H_0: $\mu_1 = \mu_2$ $\quad H_1$: $\mu_1 > \mu_2$

$$U = n_1 n_2 + \frac{n_2(n_2+1)}{2} - R_2 = 15(11) + \frac{11(12)}{2} - 115.5 = 115.5$$

$$\mu_U = \frac{n_1 n_2}{2} = \frac{15(11)}{2} = 82.5 \qquad \sigma_U = \sqrt{\frac{n_1 n_2(n_1 + n_2 + 1)}{12}} = \sqrt{\frac{15(11)(27)}{12}} = 19.27$$

We were asked if output has been reduced. If it has, we should observe a small value of R_2 and, accordingly, a large value of U. Thus an upper-tail test is appropriate.

The upper limit of the acceptance region is:

$$\mu_U + 1.28\,\sigma_U = 82.5 + 1.28(19.27) = 107.2$$

Since $U = 115.5 > 107.2$, we reject H_0. The change has reduced output significantly.

14-20

Promotion	18	21	23	15	19	26	17	18	22	20	18	21	27
Ranks	8	14	19	1.5	10	23.5	5	8	16.5	11.5	8	14	25

Regular	22	17	15	23	25	20	26	24	16	17	23	21
Ranks	16.5	5	1.5	19	22	11.5	23.5	21	3	5	19	14

$n_1 = 13$ $\quad n_2 = 12$ $\quad \alpha = .05$

$R_1 = 164$ $\quad R_2 = 161$

H_0: $\mu_1 = \mu_2$ $\quad H_1$: $\mu_1 > \mu_2$

$$U = \mathrm{n}_1\mathrm{n}_2 + \frac{n_2(n_2+1)}{2} - R_2 = 13(12) + \frac{12(13)}{2} - 161 = 73$$

$$\mu_U = \frac{n_1 n_2}{2} = \frac{12(13)}{2} = 78 \qquad \sigma_U = \sqrt{\frac{n_1 n_2(n_1 + n_2 + 1)}{12}} = \sqrt{\frac{12(13)(26)}{12}} = 18.38$$

If the promotion produced greater sales, R_2 should be low and U should be high, so an upper tail test is appropriate. The upper limit of the acceptance region is:

$$\mu_U + 1.64\sigma_U = 78 + 1.64(18.38) = 108.14$$

Since $U = 73 < 108.14$, we accept H_0. The promotion does not increase sales.

14-22

Rural	3.19	2.05	2.82	2.16	3.84	4.00	2.91	2.75	3.01	1.98	2.58	2.76	2.94
Ranks	23	4	16	7	26	29	18	13	20	3	9	14	19

Urban	3.45	3.16	2.84	2.09	2.11	3.08	3.97	3.85	3.72	2.73	2.81	2.64	1.57	1.87	2.54	2.62
Ranks	24	22	17	5	6	21	28	27	25	12	15	11	1	2	8	10

$n_1 = 13$ $\quad n_2 = 16$ $\quad \alpha = .05$

$R_1 = 201$ $\quad R_2 = 234$

H_0: $\mu_1 = \mu_2$ $\quad H_1$: $\mu_1 \neq \mu_2$

$$U = n_1 n_2 + \frac{n_1(n_1+1)}{2} - R_1 = 13(16) + \frac{13(14)}{2} - 201 = 98$$

$$\mu_U = \frac{n_1 n_2}{2} = \frac{13(16)}{2} = 104 \qquad \sigma_U = \sqrt{\frac{n_1 n_2(n_1+n_2+1)}{12}} = \sqrt{\frac{13(16)(30)}{12}} = 22.80$$

The limits of the acceptance region are:

$$\mu_U \pm 1.96\sigma_U = 104 \pm 1.96(22.80) = 104 \pm 44.69 = [59.31, 148.69]$$

Since $U = 98$, we do not reject H_0. There is no significant difference in the two groups' first-year GPAs.

14-24

$n_1 = \#$ of A's $= 26$ $\quad r = 27$

$n_2 = \#$ of B's $= 22$ $\quad \alpha = .05$

$$\mu_r = \frac{2n_1 n_2}{n_1+n_2} + 1 = \frac{2(26)(22)}{48} + 1 = 24.83$$

$$\sigma_r = \sqrt{\frac{2n_1 n_2(2n_1 n_2 - n_1 - n_2)}{(n_1+n_2)^2(n_1+n_2-1)}} = \sqrt{\frac{2(26)(22)[2(26)(22) - 26 - 22]}{(48)^2(47)}} = 3.40$$

The limits of the acceptance region are:

$$\mu_r \pm 1.96\sigma_r = 24.83 \pm 1.96(3.40) = 24.83 \pm 6.66 = (18.17, 31.49)$$

so we accept H_0. The mix is random.

14-26

$n_1 = \#$ of men $= 14$ $\quad r = 13$

$n_2 = \#$ of women $= 14$ $\quad \alpha = .05$

$$\mu_r = \frac{2n_1 n_2}{n_1+n_2} + 1 = \frac{2(1)(14)}{28} + 1 = 15$$

$$\sigma_r = \sqrt{\frac{2n_1 n_2(2n_1 n_2 - n_1 - n_2)}{(n_1+n_2)^2(n_1+n_2-1)}} = \sqrt{\frac{2(14)(14)[2(14)(14) - 14 - 14]}{(28)^2(27)}} = 2.60$$

The limits of the acceptance region are:

$$\mu_r \pm 1.96\sigma_r = 15 \pm 1.96(2.60) = 15 \pm 5.10 = (9.90, 20.10)$$

so we accept H_0. The sequence is random, as we would have expected.

14-28

$n_1 = \#$ of A's $= 15$ $\quad r = 10$

$n_2 = \#$ of B's $= 16$ $\quad \alpha = .05$

"Eyeballing" the data shows that the ages of diners are not randomly mixed, but let's see if this is verified by hard statistical analysis.

$$\mu_r = \frac{2n_1 n_2}{n_1+n_2} + 1 = \frac{2(14)(14)}{28} + 1 = 16.484$$

$$\sigma_r = \sqrt{\frac{2n_1n_2(2n_1n_2 - n_1 - n_2)}{(n_1 + n_2)^2(n_1 + n_2 - 1)}} = \sqrt{\frac{2(15)(16)[2(15)(16) - 15 - 16]}{(31)^2(30)}} = 2.734$$

If older couples eat earlier there will be too few runs, so a lower tail test is appropriate. The lower limit of the acceptance region is:

$$\mu_r - 1.64\sigma_r = 16.484 - 1.64(2.734) = 12.00$$

so we reject H_0, since $r = 10 < 12$. The pattern is not random.

14-30 a) $n_1 = 45$ $\qquad r = 9$
$n_2 = 4$ $\qquad \alpha = .01$

$$\mu_r = \frac{2n_1n_2}{n_1 + n_2} + 1 = \frac{2(45)(4)}{49} + 1 = 8.35$$

$$\sigma_r = \sqrt{\frac{2n_1n_2(2n_1n_2 - n_1 - n_2)}{(n_1 + n_2)^2(n_1 + n_2 - 1)}} = \sqrt{\frac{2(45)(4)[2(45)(4) - 45 - 4]}{(49)^2(48)}} = .99$$

The limits of the acceptance region are:

$$\mu_r \pm 2.58\sigma_r = 8.35 \pm 2.58(.99) = 8.35 \pm 2.55 = (5.80, 10.90)$$

so we accept H_0. The sample seems to be random.

b) The acceptance region is the same as before, but now $r = 2 < 5.85$, so we reject reject H_0. Of course, this was obvious by inspection.

c) Certainly. A random sample should be roughly composed of 3 times as many analyses done by machine as by hand. If $p = .75$, the probability of observing $\overline{p} \geq 45/49 = .9184$ in a sample of size 49 is very small:

$$\sigma_{\overline{p}} = \sqrt{\frac{pq}{n}} = \sqrt{\frac{(.75)(.25)}{49}} = .0619, \quad z = \frac{.9184 - .75}{.0619} = 2.72, \quad p(z \geq 2.72) = .0033.$$

Even odder is the particular sequence observed: nine 1's, a 2, nine 1's, a 2, nine 1's, a 2, nine 1's, a 2, nine 1's.

d) The test only looks at the number of runs in the sample. It <u>does not</u> see if the sample proportion is reasonable. Also, no test would pick up the exact pattern that we noted in part (c).

14-32 $n_1 = \#$ of men $= 29$ $\qquad r = 17$
$n_2 = \#$ of women $= 11$ $\qquad \alpha = .05$

$$\mu_r = \frac{2n_1n_2}{n_1 + n_2} + 1 = \frac{2(29)(11)}{40} + 1 = 16.95$$

$$\sigma_r = \sqrt{\frac{2n_1n_2(2n_1n_2 - n_1 - n_2)}{(n_1 + n_2)^2(n_1 + n_2 - 1)}} = \sqrt{\frac{2(29)(11)[2(29)(11) - 29 - 11]}{(40)^2(39)}} = 2.47$$

The limits of the acceptance region are:

$$\mu_r \pm 1.96\sigma_r = 16.95 \pm 1.96(2.47) = 16.95 \pm 4.84 = (12.11, 21.79)$$

so we accept H_0. The sequence is random, as we would have expected.

14-34

Amount of overtime	5	8	2	4	3	7	1	6
Years employed	1	6	4.5	2	7	8	4.5	3
d	4	2	−2.5	2	−4	−1	−3.5	3
d^2	16	4	6.25	4	16	1	12.25	9

$\sum d^2 = 68.5$ $\qquad n = 8$ $\qquad \alpha = .01$

$H_0\colon\ \rho_s = 0$ $\qquad H_1\colon\ \rho_s \neq 0$

$$r_s = 1 - \frac{6\sum d^2}{n(n^2-1)} = 1 - \frac{6(68.5)}{8(63)} = .1845$$

From Appendix Table 7, the critical values for r_s are $\pm$ .8571.
Since .1845 < .8571, we accept H_0. The correlation is not significant.

14-36

Company Ranks of:	A	B	C	D	E	F	G	H	I	J	K
Expenses	1	7	8	11	10	5	6	3	4	2	9
Accidents	10.5	4.5	6	1	4.5	7.5	2.5	10.5	7.5	9	2.5
d	−9.5	2.5	2	10	5.5	−2.5	3.5	−7.5	−3.5	−7	6.5
d^2	90.25	6.25	4	100	30.25	6.25	12.25	56.25	12.25	49	42.25

$\sum d^2 = 409$ $\qquad n = 11$ $\qquad \alpha = .01$

$H_0\colon\ \rho_s = 0$ $\qquad H_1\colon\ \rho_s \neq 0$

$$r_s = 1 - \frac{6\sum d^2}{n(n^2-1)} = 1 - \frac{6(409)}{11(120)} = -.8591$$

From Appendix Table 7, the critical values for r_s are $\pm$ .7455. Since −.8591 < −.7455, we reject H_0. There is a significant correlation.

14-38

Applicant	1	2	3	4	5	6	7	8	9	10	11	12	13	14
Interview 1	1	11	13	2	12	10	3	4	14	5	6	9	7	8
Interview 2	4	12	11	2	14	10	1	3	13	8	6	7	9	5
d	−3	−1	2	0	−2	0	2	1	1	−3	0	2	−2	3
d^2	9	1	4	0	4	0	4	1	1	9	0	4	4	9

$\sum d^2 = 50$ $\qquad n = 14$ $\qquad \alpha = .01$

$H_0\colon\ \rho_s = 0$ $\qquad H_1\colon\ \rho_s > 0$

$$r_s = 1 - \frac{6\sum d^2}{n(n^2-1)} = 1 - \frac{6(50)}{14(195)} = .8901$$

Critical r_s (one tail) = .6220 < .8901, so we can conclude that the rankings are positively correlated.

14-40

Individual	Interview Score	Resume Score	d	d^2
1	15	21	−6	36
2	25	9	16	256
3	1	4	−3	9
4	18	27	−9	81
5	11.5	15	−3.5	12.25
6	30	33	−3	9
7	4	11	−7	49
8	23	30	−7	49
9	34.5	6	28.5	812.25
10	10	10	0	0
11	2	7.5	−5.5	30.25
12	20.5	13.5	7	49
13	30	26	4	16
14	5	17.5	12.5	156.25
15	26.5	12	14.5	210.25

16	7	1	6	36
17	23	13.5	9.5	90.25
18	6	16	−10	100
19	15	20	−5	25
20	19	24	−5	25
21	17	7.5	9.5	90.25
22	26.5	5	21.5	462.25
23	3	3	0	0
24	11.5	17.5	−6	36
25	9	2	7	49
26	13	24	−11	121
27	8	19	−11	121
28	34.5	24	10.5	110.25
29	15	32	−17	289
30	23	28	−5	25
31	30	29	1	1
32	20.5	34	−13.5	182.25
33	28	22	6	36
34	32.5	35	−2.5	6.25
35	32.5	31	1.5	2.25

$\sum d^2 = 3583$

$$r_s = 1 - \frac{6\sum d^2}{n(n^2-1)} = 1 - \frac{6(3583)}{35(35^2-1)} = .4982$$

For a one-tail test at the .01 significance level, the critical z value is 2.33.

$$\sigma_{r_s} = \frac{1}{\sqrt{n-1}} = \frac{1}{\sqrt{34}} = .1715$$

H_0: $\rho_s = 0$ H_1: $\rho_s > 0$

Thus, the critical value for $r_s = 2.33(.1715) = .3996$. Since $.4982 > .3996$, the firm should recommend that the personal interviews no longer be used.

14-42

Interval	1000	1200	900	1450	2000	1300	1650	1700	500	2100
Rank	8	7	9	5	2	6	4	3	10	1
Repair time	40	54	41	60	65	50	42	65	43	66
Rank	10	5	9	4	2.5	6	8	2.5	7	1
d	−2	2	0	1	−0.5	0	−4	0.5	3	0
d^2	4	4	0	1	0.25	0	16	0.25	9	0

$\sum d^2 = 34.5$ $n = 10$ $\alpha = .10$

H_0: $\rho_s = 0$ H_1: $\rho_s \neq 0$

$$r_s = 1 - \frac{6\sum d^2}{n(n^2-1)} = 1 - \frac{6(34.5)}{10(99)} = .7909$$

From Appendix Table 7, the critical values for r_s are $\pm .5515$. Since $.7909 > .5515$, we reject H_0. The rank correlation is significant.

14-44 a) With $\mu = 98.6$ and $\sigma = 3.78$,

$$P(x < 92.0) = P\left(z < \frac{92.0-98.6}{3.78}\right) = P(z < -1.75) = 0.5-0.4599 = 0.0401$$

$$P(92.0 \le x < 96.0) = P\left(-1.75 \le z < \frac{96.0-98.6}{3.78}\right)$$
$$= P(-1.75 \le z < -0.69) = 0.4599-0.2549 = 0.2050$$

$$P(96.0 \le x < 100.0) = P\left(-0.69 \le z < \frac{100.0-98.6}{3.78}\right)$$
$$= P(-0.69 \le z < 0.37) = 0.2549 + 0.1443 = 0.3992$$
$$P(100.0 \le x < 104.0) = P\left(0.37 \le z < \frac{104.0-98.6}{3.78}\right)$$
$$= P(0.37 \le z < 1.43) = 0.4236 - 0.1443 = 0.2793$$

$$P(104.0 \le x) = P(1.43 \le z) = 0.5 - 0.4236 = .0764$$

b, c) $n = 69 + 408 + 842 + 621 + 137 = 2077$

For each interval, f_e is 2077 times the probability found in part (a).

Interval	f_o	cum. f_o	F_o	f_e	cum. f_e	F_e	$\lvert F_e - F_o \rvert$
< 92	69	69	0.0332	83.29	83.29	0.0401	0.0069
92 - 95.99	408	477	0.2297	425.78	509.07	0.2451	0.0154 ←
96 - 99.99	842	1319	0.6351	829.14	1338.21	0.6443	0.0092
100 - 103.99	621	1940	0.9340	580.11	1918.32	0.9236	0.0104
≥ 104	137	2077	1.0000	158.68	2077.00	1.0000	0.0000

$D_n = 0.0154$; $D_{table} = \frac{1.22}{\sqrt{n}} = \frac{1.22}{\sqrt{2077}} = 0.0268$; $D_n < D_{table}$, so do not reject Ho. The data are described by a normal distribution, with $\mu = 98.6$ and $\sigma = 3.78$.

14-46

Class	f_o	cum. f_o	F_o	f_e	cum. f_e	F_e	$\lvert F_e - F_o \rvert$
25-30	9	9	.072	6	6	.048	.024
31-36	22	31	.248	17	23	.184	.064 ←
37-42	25	56	.448	32	55	.440	.008
43-48	30	86	.688	35	90	.720	.032
49-54	21	107	.856	18	108	.864	.008
55-60	12	119	.952	13	121	.968	.004
61-66	6	125	1.000	4	125	1.000	.000

$D_n = .064$; $D_{table} = \frac{1.22}{\sqrt{n}} = \frac{1.22}{\sqrt{125}} = .1091$; $D_n < D_{table}$, so accept H_0.
The data are well described by the suggested distribution.

14-48 $\lambda = 1$, $e^{-\lambda} = .367879$

x	f_o	cum. f_o	F_o	F_e	$\lvert F_e - F_o \rvert$
0	25	25	.1250	.3679	.2429
1	45	70	.3500	.7358	.3858 ←
2	67	137	.6850	.9197	.2347
3	43	180	.9000	.9810	.0810
≥ 4	20	200	1.0000	1.0000	.0000

$D_n = .3858$; $D_{table} = \frac{1.36}{\sqrt{n}} = \frac{1.36}{\sqrt{200}} = .0962$; $D_n > D_{table}$, so reject H_0.

The data are not well described by a Poisson distribution with $\lambda = 1$.

14-50

Before Data	27	15	20	24	13	18	30	46	15	29	17	21	18
Ranks	22	7.5	15.5	19	5	12.5	25	26	7.5	24	11	17	12.5

After Data	26	23	19	12	25	9	16	12	28	20	16	14	11
Ranks	21	18	14	3.5	20	1	9.5	3.5	23	15.5	9.5	6	2

$n_1 = 13$ $\quad n_2 = 13$ $\quad \alpha = .02$
$R_1 = 204.5$ $\quad R_2 = 146.5$

H_0: $\mu_1 = \mu_2$ $\qquad$ H_1: $\mu_1 > \mu_2$

$$U = n_1 n_2 + \frac{n_2(n_2+1)}{2} - R_2 = 13(13) + \frac{13(14)}{2} - 146.5 = 113.5$$

$$\mu_U = \frac{n_1 n_2}{2} = \frac{13(13)}{2} = 84.5 \qquad \sigma_U = \sqrt{\frac{n_1 n_2(n_1+n_2+1)}{12}} = \sqrt{\frac{13(13)(27)}{12}} = 19.52$$

If complaints declined, R_2 should be small and hence U should be large. An upper-tail test is appropriate. The upper limit of the acceptance region is:

$$\mu_U + 2.05\sigma_U = 84.5 + 2.05(19.5) = 124.47$$

so we accept H_0, since $U = 113.5 < 134.47$. There has not been a significant reduction.

14-52

Infantry	72	80	86	90	95	92	88	96	91	82
Ranks	2	6.5	13	15.5	19	18	14	20	17	9.5

Transport	80	79	90	82	81	84	78	74	85	71
Ranks	6.5	5	15.5	9.5	8	11	4	3	12	1

$n_1 = 10$ $\qquad$ $n_2 = 10$ $\qquad$ $\alpha = .05$

$R_1 = 134.5$ $\qquad$ $R_2 = 75.5$

H_0: $\mu_1 = \mu_2$ $\qquad$ H_1: $\mu_1 > \mu_2$

$$U = n_1 n_2 + \frac{n_2(n_2+1)}{2} - R_2 = 10(10) + \frac{10(11)}{2} - 75.5 = 79.5$$

$$\mu_U = \frac{n_1 n_2}{2} = \frac{10(10)}{2} = 50 \qquad \sigma_U = \sqrt{\frac{n_1 n_2(n_1+n_2+1)}{12}} = \sqrt{\frac{10(10)(21)}{12}} = 13.23$$

If ratings are lower in the transport command, R_2 should be low and hence U should be high, so an upper-tail test is appropriate. The upper limit of the acceptance region is:

$$\mu_U + 1.64\sigma_U = 50 + 1.64(13.23) = 71.70$$

so we reject H_0, since $U = 79.5 > 71.70$. The transport command has significantly lower ratings.

14-54

Bus. Wk.	1	2	3	4	5	6	7	8	9	10	11	12	13	14	15	16	17	18	19	20
USN&WR	4	6	2	3	7	10	1	18	8	16	11	9	5	12	17	14	15	13	19	20
Sign	+	+	−	−	+	+	−	+	−	−	0	−	−	−	+	−	−	−	0	0

20 responses: 6(+); 11(−); 3(0)

$$\bar{p} = \frac{6}{17} = .3529 \qquad \sigma_{\bar{p}} = \sqrt{\frac{pq}{n}} = \sqrt{\frac{.5(.5)}{17}} = 0.1213$$

H_0: $p = .5$ (no difference) $\qquad$ H_1: $p \neq .5$

At $\alpha = .10$, the limits of the acceptance region are:

$$.5 \pm 1.645\sigma_{\bar{p}} = .5 \pm 1.645(.1213) = [.3005, .6995]$$

so we accept H_0, since $.3011 < .3529 < .6995$. There is no significant difference between the magazine rankings.

14-56 Although historical data enable him to know what sort of weather to expect at any season of the year, the weather conditions that actually occur on any given day are quite random.

14-58

Letters	35	85	90	92	88	46	78	57	85	67
Ranks	1	10.5	18	19	16	4	11	5	13.5	7
Brochures	42	74	82	87	45	73	89	75	60	94
Ranks	2	9	12	15	3	8	17	10	6	20

$n_1 = 10$ $\qquad n_2 = 10$ $\qquad \alpha = .15$ $\qquad R_1 = 108$ $\qquad R_2 = 102$

H_0: $\mu_1 = \mu_2$ $\qquad H_1$: $\mu_1 < \mu_2$

$$U = n_1 n_2 + \frac{n_2(n_2+1)}{2} - R_2 = 10(10) + \frac{10(11)}{2} - 102 = 53$$

$$\mu_U = \frac{n_1 n_2}{2} = \frac{10(10)}{2} = 50 \qquad \sigma_U = \sqrt{\frac{n_1 n_2 (n_1 + n_2 + 1)}{12}} = \sqrt{\frac{10(10)(21)}{12}} = 13.23$$

If brochures are more effective, then R_2 should be high, and hence U should be low, so a lower-tail test is appropriate. Since $U > \mu_U$, we must accept H_0. Her hunch is not supported by the data.

14-60

Size	R	5-yr. Avg.	R	d	d^2	1992	R	d	d^2
21.05	1	11.24	5	−4	16	9.51	15	−10	100
14.03	2	9.50	10	−8	64	11.08	9	1	1
9.48	3	8.99	14	−11	121	11.35	8	6	36
8.23	4	7.00	17	−13	169	9.53	14	3	9
5.77	5	8.73	15	−10	100	10.87	10	5	25
5.64	6	11.57	4	2	4	16.33	1	3	9
5.62	7	9.38	11	−4	16	15.11	2	9	81
5.10	8	9.34	12	−4	16	11.44	6	6	36
4.98	9	11.07	6	3	9	5.77	18	−12	144
4.80	10	9.59	8	2	4	14.71	3	5	25
4.67	11	10.03	7	4	16	11.42	7	0	0
4.66	12	14.70	1	11	121	8.55	16	−15	225
4.65	13	7.29	16	−3	9	12.45	4	12	144
4.60	14	9.06	13	1	1	11.59	5	8	64
4.47	15	6.25	18	−3	9	2.02	19	−1	1
4.40	16	9.52	9	7	49	10.84	11	−2	4
4.29	17	11.80	3	14	144	10.51	12	−9	81
4.02	18	5.47	19	−1	1	7.00	17	2	4
4.01	19	14.55	2	17	289	1.24	20	−18	324
3.97	20	4.78	20	0	0	9.92	13	7	49
					$\sum d^2 = 1210$				$\sum d^2 = 1362$

a) H_0: $\rho_s = 0$ $\qquad H_1$: $\rho_s \neq 0$

$$r_s = 1 - \frac{6\sum d^2}{n(n^2-1)} = 1 - \frac{6(1210)}{20(399)} = 0.0902$$

From Appendix Table 7, the critical values for r_s are $\pm$ 0.2977. Since the probability value is greater than 0.20, we accept H_0. There is not a significant relationship between fund size and the annualized 5 year total return.

b) H_0: $\rho_s = 0$ $\qquad H_1$: $\rho_s \neq 0$

$$r_s = 1 - \frac{6\sum d^2}{n(n^2-1)} = 1 - \frac{6(1362)}{20(399)} = -0.0241$$

So again we accept H_0. There is not a significant relationship between the 1992 total return and the annualized 5 year return.

14-62

Graphite	110	120	130	110	100	105	110	130	145	125
Ranks	9.5	13.5	16.5	9.5	2.5	6	9.5	16.5	20	15

Bronze	100	110	135	105	105	100	100	115	135	120
Ranks	2.5	9.5	18.5	6	6	2.5	2.5	12	18.5	13.5

$n_1 = 10$ $\quad n_2 = 10$ $\quad \alpha = .05$

$R_1 = 118.5$ $\quad R_2 = 91.5$

H_0: $\mu_1 = \mu_2$ $\quad H_1$: $\mu_1 \neq \mu_2$

$$U = n_1 n_2 + \frac{n_2(n_2+1)}{2} - R_2 = 10(10) + \frac{10(11)}{2} - 91.5 = 63.5$$

$$\mu_U = \frac{n_1 n_2}{2} = \frac{10(10)}{2} = 50 \qquad \sigma_U = \sqrt{\frac{n_1 n_2(n_1 + n_2 + 1)}{12}} = \sqrt{\frac{10(10)(21)}{12}} = 13.23$$

The limits of the acceptance region are:

$$\mu_U \pm 1.96\sigma_U = 50 \pm 1.96(13.23) = 50 \pm 25.93 = (24.07, 75.93)$$

Since $U = 63.5$, we accept H_0. The stopping distances are not significantly different.

14-64

Raw Score	63	59	50	60	66	57	76	81	58	65
Rank	8	5	1	6	10	3	14	17	4	9
Interview #	1	2	3	4	5	6	7	8	9	10
d	7	3	−2	2	5	−3	7	9	−5	1
d^2	49	9	4	4	25	9	49	81	25	1

Raw Score	77	61	53	74	82	70	75	90	80	89
Rank	15	7	2	12	18	11	13	20	16	19
Interview #	11	12	13	14	15	16	17	18	19	20
d	4	−5	−11	−2	3	−5	−4	2	−3	−1
d^2	16	25	121	4	9	25	16	4	9	1

$\sum d^2 = 486$ $\quad n = 20$ $\quad \alpha = .02$

H_0: $\rho_s = 0$ $\quad H_1$: $\rho_s \neq 0$

$$r_s = 1 - \frac{6\sum d^2}{n(n^2-1)} = 1 - \frac{6(486)}{20(399)} = .6346$$

From Appendix Table 7, the critical values for r_s are $\pm$ .5203, so we reject H_0, since .6346 > .5203. This supports her suspicion.

14-66

x	f_o	cum. f_o	F_o	F_e	$\lvert F_e - F_o \rvert$
0	5	5	.0556	.1785	.1229 ←
1	35	40	.4444	.5630	.1186
2	30	70	.7778	.8735	.0957
3	13	83	.9222	.9850	.0628
4	7	90	1.0000	1.0000	.0000

$D_n = .1229$; $D_{table} = \frac{1.36}{\sqrt{n}} = \frac{1.36}{\sqrt{90}} = .1434$; $\quad D_n < D_{table}$, so accept H_0.

The data are well described by a binomial distribution with $n = 4$ and $p = .35$.

14-68

Bulk	Ranks	Tanker	Ranks	PPC	Ranks
1978	27	1974	12	1983	43
1978	27	1973	8.5	1982	39.5
1982	39.5	1977	23.5	1969	3
1982	39.5	1977	23.5	1968	1.5
1975	17	1978	27	1968	1.5
1975	17	1977	23.5	1986	48.5
1990	63	1971	5	1986	48.5
1990	63	1970	4	1986	48.5
1973	8.5	1973	8.5	1987	51.5
1973	8.5	1975	17	1989	57.5
1981	35.5	1974	12	1988	53
1983	43	1974	12	1989	57.5
1983	43	1989	57.5	1989	57.5
1989	57.5	1990	63	1979	29.5
1989	57.5	1972	6	1981	35.5
1980	32	1989	57.5	1981	35.5
1980	32	1989	57.5	1982	39.5
1977	23.5	1976	21		
1975	17	1975	17		
1975	17	1975	17		
1985	45.5	1986	48.5		
1985	45.5	1987	51.5		
		1980	32		
		1981	35.5		
		1979	29.5		

$n_1 = 22$ $n_2 = 25$ $n_3 = 17$ $R_1 = 759$ $R_2 = 670$ $R_3 = 651$

H_0: $\mu_1 = \mu_2 = \mu_3$ H_1: the μ's are not all the same

$$K = \frac{12}{n(n+1)}\sum \frac{R_j^2}{n_j} - 3(n+1) = \frac{12}{64(65)}\left(\frac{759^2}{22} + \frac{670^2}{25} + \frac{651^2}{17}\right) - 3(65) = 4.2432$$

With $3 - 1 = 2$ degrees of freedom, the upper limit of the acceptance region is $\chi^2_{2,.10} = 4.605$. The probability value is > .10, so we accept H_0. The average ages of the three types of vessels are not significantly different.

14-70

Region	US	Latin America	Africa	Europe	USSR	India	China
Population rank	1	3	5	4	2	6	7
Energy rank	7	3	2	6	5	1	4
d	−6	0	3	−2	−3	5	3
d^2	36	0	9	4	9	25	9

$\sum d^2 = 92$ $n = 10$ $\alpha = .10$

H_0: $\rho_s = 0$ H_1: $\rho_s < 0$

$$r_s = 1 - \frac{6\sum d^2}{n(n^2-1)} = 1 - \frac{6(92)}{7(48)} = -.6429$$

From Appendix Table 7, the critical value for r_s is −.5357, so we reject H_0, since −.6429 < −.5357. SavEnergy's claim is supported by the data.

14-72

American	46	3	7	21	10	10	1	6	7	17	7	16	6	1	0	10
National	22	0	5	19	7	2	1	4	5	14	3	18	4	4	1	4
Sign	−	−	−	−	−	−	0	−	−	−	−	+	−	+	+	−

16 responses: 3(+); 12(−); 1(0)

H_0: $p = .5$ H_1: $p > .5$ (Here p is the probability of getting a −.)

For $n = 15$ and $p = .5$, the probability of $\geq$ 12 −'s is .0176 (Appendix Table 3). Since .0176 < .05, we reject H_0; American League players do suffer more injuries.

14-74 a) $\frac{1500}{9000} = 0.1667$ 0.1667(42) = 7 in each group

b)

f_o	f_e	$f_o - f_e$	$(f_o - f_e)^2$	$\frac{(f_o - f_e)^2}{f_e}$
10	7	3	9	1.2857
9	7	2	4	0.5714
9	7	2	4	0.0373
1	7	−6	36	5.1429
9	7	2	4	0.5714
4	7	−3	9	1.2857

$$\chi^2 = \sum \frac{(f_o - f_e)}{f_e} = 9.4285$$

$\chi^2_{.10,5} = 9.236$

Since $\chi^2 = 9.4285$, we reject the H_0. There is reason to believe that the called bonds were not selected randomly.

c)

Class	f_o	cum. f_o	F_o	f_e	cum. f_e	F_e	$\lvert F_e - F_o \rvert$
1-1500	10	10	.2381	7	7	.1667	.0714
1501-3000	9	19	.4524	7	14	.3333	.1191
3001-4500	9	28	.6667	7	21	.5000	.1667←
4501-6000	1	29	.6905	7	28	.6667	.0238
6001-7500	9	38	.9048	7	35	.8333	.0715
7501-9000	4	42	1.0000	7	42	1.0000	.0000

$D_n = .1667$; $D_{\text{table}} = \frac{1.14}{\sqrt{n}} = \frac{1.14}{\sqrt{42}} = .1759$

Since the probability value is > .15 ($D_n < D_{\text{table}}$), we accept H_0. The called bonds appear to have been selected randomly.

d) The Kolmogorov-Smirnov test suggested that the bonds were selected randomly, whereas the χ^2 test indicated the opposite result (at α=.10). Since the χ^2 test was barely significant (.10) and the K-S test is more powerful, we should conclude that the bonds were randomly selected.

14-76 $\lambda = 6,\ e^{-\lambda} = .002479$

x	f_0	cum. f_0	F_0	F_e	$\lvert F_e - F_0 \rvert$
0	0	0	.0000	.0025	.0025
1	5	5	.1000	.0174	.0826
2	3	8	.1600	.0620	.0980
3	2	10	.2000	.1512	.0488
4	6	16	.3200	.2851	.0349
5	6	22	.4400	.4457	.0057
6	2	24	.4800	.6063	.1263
7	6	30	.6000	.7440	.1440 ←
8	10	40	.8000	.8473	.0473
9	4	44	.8800	.9161	.0361
10	4	48	.9600	.9574	.0026
≥ 11	2	50	1.0000	1.0000	.0000

$D_n = .1440$; $D_{table} = 1.36/\sqrt{n} = 1.36/\sqrt{50} = .1923$; $D_n < D_{table}$, so accept H_0. The data are well described by a Poisson distribution with $\lambda = 6$.

14-78

A rank	B rank	d=A−B	d^2	C rank	d=A−C	d^2	D rank	d=A−D	d^2
1	2	-1	1	5	-4	16	3	-2	4
2	4	-2	4	3	-1	1	4	-2	4
3	6	-3	9	7	-4	16	5	-2	4
4	3	1	1	2	2	4	1	3	9
5	10	-5	25	11	-6	36	9	-4	16
6	5	1	1	6	0	0	2	4	16
7	22	-15	225	17	-10	100	24	-17	289
8	9	-1	1	4	4	16	15	-7	49
9	18	-9	81	19	-10	100	19	-10	100
10	19	-9	81	30	-20	400	17	-7	49
11	20	-9	81	9	2	4	22	-11	121
12	7	5	25	15	-3	9	11	1	1
13	12	1	1	14	-1	1	13	0	0
14	25	-11	121	12	2	4	28	-14	196
15	16	-1	1	24	-9	81	14	1	1
16	17	-1	1	16	0	0	18	-2	4
17	8	9	81	8	9	81	7	10	100
18	11	7	49	10	8	64	6	12	144
19	24	-5	25	23	-4	16	20	-1	1
20	1	19	361	1	19	361	8	12	144
21	26	-5	25	20	1	1	30	-9	81
22	29	-7	49	27	-5	25	27	-5	25
23	13	10	100	18	5	25	10	13	169
24	28	-4	16	21	3	9	26	-2	4
25	30	-5	25	29	-4	16	29	-4	16
26	14	12	144	13	13	169	23	3	9
27	23	4	16	26	1	1	21	6	36
28	27	1	1	28	0	0	25	3	9
29	15	14	196	22	7	49	12	17	289
30	21	9	81	25	5	25	16	14	196
		$\Sigma d^2 =$	1828			1630			2086

$n = 30$

$$r_s = 1 - \frac{6\sum d^2}{n(n^2-1)} = 1 - \frac{6(1828)}{30(899)} = .5933 \text{ for the League of Women Voters}$$

$$= 1 - \frac{6(1630)}{30(899)} = .6374 \text{ for college students}$$

$$= 1 - \frac{6(2086)}{30(899)} = .5359 \text{ for civic club members}$$

The college students seem to have the most accurate perception of the risks. However, we do not know if the observed differences in rank correlation coefficients are significant, because we have not covered tests for comparing rank correlations.

14-80

1995 Rank	1	2	3	4	5	6	7	8	9	10
1996 Rank	1	2	8	5	4	10	3	7	6	9
d	0	0	−5	−1	1	−4	4	1	3	1
d^2	0	0	25	1	1	16	16	1	9	1

$\sum d^2 = 70$ $\quad n = 10$ $\quad \alpha = .10$

H_0: $\rho_s = 0$ $\quad H_1$: $\rho_s \neq 0$

$$r_s = 1 - \frac{6\sum d^2}{n(n^2-1)} = 1 - \frac{6(70)}{10(99)} = .5758$$

A two-tailed test is appropriate, since we want to see if the rankings have changed. From Appendix Table 7, the critical values for r_s are ± 0.5515, so we reject H_0. The rankings have not changed significantly.

CHAPTER 15

TIME SERIES

15-2 To determine what patterns exist within the data over the period examined.

15-4 Demands for services such as water and sewer would perhaps not be met; adjustment of the tax rate to provide for municipal services might lag behind the actual demand for those services. Extra resources would likely be needed to allow a smooth municipal operation in a situation where forecasting is so poor.

15-6 Seasonal

15-8 Cyclical variation

15-10 Secular trend

15-12 a)

Year	x	Y	xY	x^2
1986	−5	6.4	−32.0	25
1987	−4	11.3	−45.2	16
1988	−3	14.7	−44.1	9
1989	−2	18.4	−36.8	4
1990	−1	19.6	−19.6	1
1991	0	25.7	0.0	0
1992	1	32.5	32.5	1
1993	2	48.7	97.4	4
1994	3	55.4	166.2	9
1995	4	75.7	302.8	16
1996	5	94.3	471.5	25
	0	402.7	892.7	110

$a = \overline{Y} = \frac{402.7}{11} = 36.6091$ $\qquad b = \frac{\sum xY}{\sum x^2} = \frac{892.7}{110} = 8.1155$

$\widehat{Y} = 36.6091 + 8.1155x$ (where 1991 = 0 and x units = 1 year)

b) 1997: $\widehat{Y} = 36.6091 + 8.1155(6) = 85.3$ homes
1998: $\widehat{Y} = 36.6091 + 8.1155(7) = 93.4$ homes
1999: $\widehat{Y} = 36.6091 + 8.1155(8) = 101.5$ homes

15-14

Year	x	Y	xY	x^2	x^2Y	x^4
1989	−7	82.4	−576.8	49	4037.6	2401
1990	−5	125.7	−628.5	25	3142.5	625
1991	−3	276.9	−830.7	9	2492.1	81
1992	−1	342.5	−342.5	1	342.5	1
1993	1	543.6	543.6	1	543.6	1
1994	3	691.5	2074.5	9	6223.5	81
1995	5	782.4	3912.0	25	19560.0	625
1996	7	889.5	6226.5	49	43585.5	2401
	0	3734.5	10378.1	168	79927.3	6216

a) $a = \overline{Y} = \frac{3734.5}{8} = 466.8125 \qquad b = \frac{\sum xY}{\sum x^2} = \frac{10378.1}{168} = 61.7744$

$\widehat{Y} = 466.8125 + 61.7744x$ (where 1992.5 = 0 and x units = .5 year)

b) Equations 15.6 and 15.7 become:

$$\sum Y = na + c\sum x^2 \qquad 3734.5 = 8a + 168c$$

$$\sum x^2 Y = a\sum x^2 + c\sum x^4 \qquad 79927.3 = 168a + 6216c$$

Solving these simultaneously, we get:

$a = 455.0719,\ c = 0.5591$

$\widehat{Y} = 455.0719 + 61.7744x + 0.5591x^2$

c) Linear forecast: $\widehat{Y} = 466.8125 + 61.7744(11) = 1146.33$ thousand mice

Quadratic forecast: $\widehat{Y} = 455.0719 + 61.7744(11) + 0.5591(121) = 1202.24$ thousand mice

15-16

Year	x	Y	xY	x^2	x^2Y	x^4
1968	−7	5	−35	49	245	2401
1970	−6	5	−30	36	180	1296
1972	−5	8	−40	25	200	625
1974	−4	8	−32	16	128	256
1976	−3	10	−30	9	90	81
1978	−2	13	−26	4	52	16
1980	−1	15	−15	1	15	1
1982	0	18	0	0	0	0
1984	1	20	20	1	20	1
1986	2	22	44	4	88	16
1988	3	25	75	9	225	81
1990	4	25	100	16	400	256
1992	5	29	145	25	725	625
1994	6	29	174	36	1044	1296
1996	7	32	224	49	1568	2401
	0	264	574	280	4980	9352

a) $a = \overline{Y} = \frac{264}{15} = 17.6 \qquad b = \frac{\sum xY}{\sum x^2} = \frac{574}{280} = 2.05$

$\widehat{Y} = 17.6 + 2.05x$ (where 1982 = 0 and x units = two years)

b) Equations 15.6 and 15.7 become:

$$\sum Y = na + c\sum x^2 \qquad 264 = 15a + 280c$$

$$\sum x^2 Y = a\sum x^2 + c\sum x^4 \qquad 4980 = 280a + 9352c$$

Solving these simultaneously, we get:

$a = 17.3647,\ c = 0.0126$

$\widehat{Y} = 17.3647 + 2.0500x + 0.0126x^2$

c) Political resistance to increased rates makes it unlikely that the quadratic trend would continue to be a good predictor.

15-18 a) Since the rate of increase in the pollution rating is itself increasing, a second degree trend would fit the data better than a linear trend.

b) However, as the air gets more polluted and citizens get more concerned, actions will be taken to control pollution, so the predictions of the second-degree tend will in all likelihood be too dire.

c) Since public or political action will likely reduce pollution, neither a linear nor a second-degree estimating equation will be accurate.

15-20

a)

Year	1989	1990	1991	1992	1993	1994	1995
Y	1.100	1.500	1.900	2.100	2.400	2.900	3.500
$\widehat{Y}$	1.174	1.449	1.764	2.119	2.514	2.949	3.424
$100\,Y/\widehat{Y}$	93.70	103.52	107.71	99.100	95.47	98.34	102.22
b) $100(Y/\widehat{Y}-1)$	−6.30	3.52	7.71	−0.90	−4.53	1.66	2.22

c)

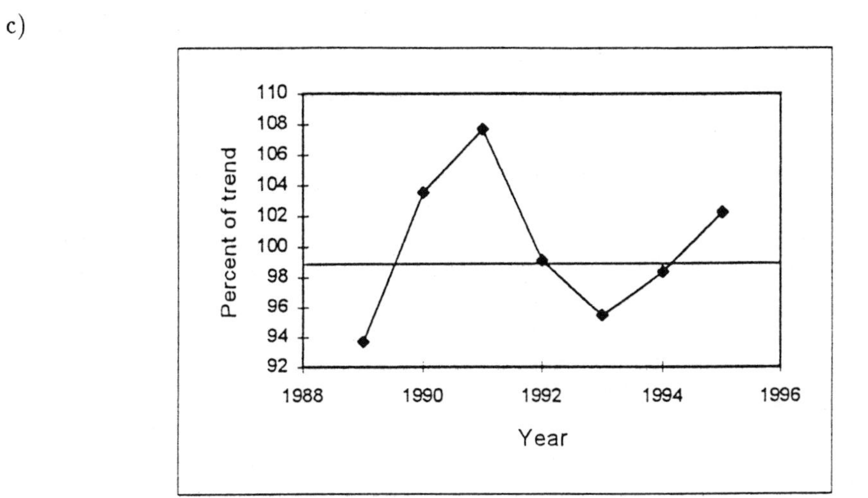

d) Largest fluctuation (by both methods) was in 1991.

15-22

a)

Year	1991	1992	1993	1994	1995
Y	32.00	46.00	50.00	66.00	68.00
$\widehat{Y}$	34.00	43.20	52.40	61.60	70.80
$100\,Y/\widehat{Y}$	94.12	106.48	95.42	107.14	96.05
b) $100(Y/\widehat{Y}-1)$	−5.88	6.48	−4.58	7.14	−3.95

c)

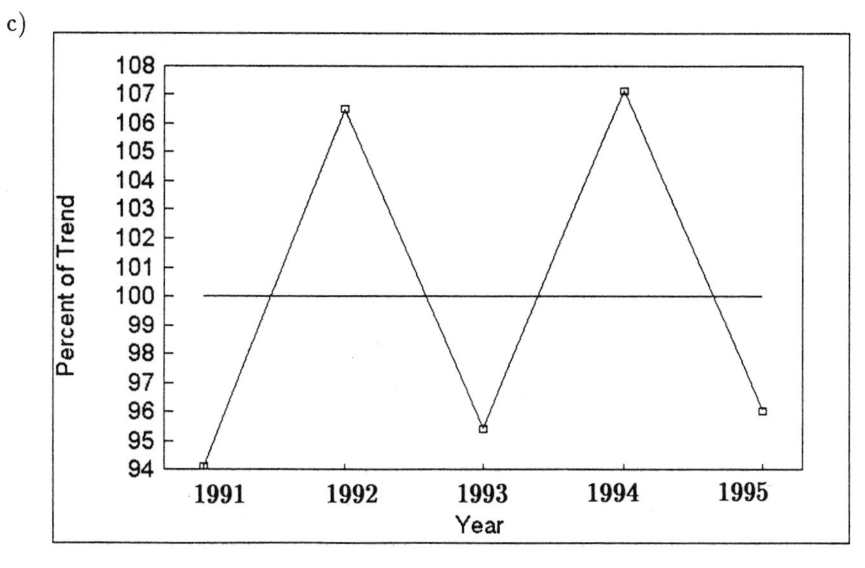

d) Largest fluctuation (by both methods) was in 1994.

15-24

Year	x	Y	xY	x^2	$\widehat{Y}$	$100(Y/\widehat{Y})$	$100(Y/\widehat{Y}-1)$
1989	−3	21.0	−63.0	9	21.2643	98.76	−1.24
1990	−2	19.4	−38.8	4	22.3000	87.00	−13.00
1991	−1	22.6	−22.6	1	23.3357	96.85	−3.15
1992	0	28.2	0.0	0	24.3714	115.71	15.71
1993	1	30.4	30.4	1	25.4071	119.65	19.65
1994	2	24.0	48.0	4	26.4428	90.76	−9.24
1995	3	25.0	75.0	9	27.4785	90.98	−9.02
	0	170.6	29.0	28			

a) $a = \overline{Y} = \frac{170.6}{7} = 24.3714$ $\qquad b = \frac{\sum xY}{\sum x^2} = \frac{29}{28} = 1.0357$

$\widehat{Y} = 24.3714 + 1.0357x$ (where 1992 = 0 and x units = 1 year)

b) See the next-to-the-last column above.

c) See the last column above.

d) Largest fluctuation (by both methods) was in 1993.

15-26 a, b)

Year	Quarter	Actual Receivables	4–Quarter Moving Average	Centered Moving Average	Ratio of Actual to CMA
1991	Spring	102			
	Summer	120	97.50		
	Fall	90	99.50	98.500	0.9137
	Winter	78	101.00	100.250	0.7781
1992	Spring	110	102.25	101.625	1.0824
	Summer	126	103.50	102.875	1.2248
	Fall	95	103.75	103.625	0.9168
	Winter	83	104.25	104.000	0.7981
1993	Spring	111	104.75	104.500	1.0622
	Summer	128	105.50	105.125	1.2176
	Fall	97	106.50	106.000	0.9151
	Winter	86	108.25	107.375	0.8009
1994	Spring	115	109.75	109.000	1.0550
	Summer	135	111.00	110.375	1.2231
	Fall	103	112.75	111.875	0.9207
	Winter	91	115.00	113.875	0.7991
1995	Spring	122	116.75	115.875	1.0529
	Summer	144	118.50	117.625	1.2242
	Fall	110			
	Winter	98			

c)

Year	Spring	Summer	Fall	Winter	
1991			.9137	.7781	
1992	1.0824	1.2248	.9168	.7981	
1993	1.0622	1.2176	.9151	.8009	
1994	1.0550	1.2231	.9207	.7991	
1995	1.0529	1.2242			
Modified sum	2.1172	2.4473	1.8319	1.5972	
Modified mean	1.0586 +	1.2236 +	.9159 +	.7986	= 3.9967/4 = .999175
Seasonal index	1.0595	1.2246	.9167	.7993	

15-28

Year	Baseball		Football		Basketball		Hockey
1992	96		128		116		77
1993	92		131		126		69
1994	94		113		117		84
1995	97		118		126		89
1996	91		121		124		81
Modified sum	279		367		366		242
Modified mean	93.00	+	122.33	+	122.00	+	80.67 = 418.00/4 = 104.5
Seasonal index	89.00		117.06		116.75		77.20

15-30 a, b)

Year	Quarter	Actual Enrollment	4-Quarter Moving Average	Centered Moving Average	Percentage of Actual to Moving Average
1991	Fall	220			
	Winter	203			
	Spring	193	175.00	176.875	109.117
	Summer	84	178.75	179.375	46.829
1992	Fall	235	180.00	181.625	129.387
	Winter	208	183.25	182.250	114.129
	Spring	206	181.25	181.375	113.577
	Summer	76	181.50	181.250	41.931
1993	Fall	236	181.00	181.375	130.117
	Winter	206	181.75	181.375	113.577
	Spring	209	181.00	181.625	115.072
	Summer	73	182.25	183.375	39.809
1994	Fall	241	184.50	184.125	130.889
	Winter	215	183.75	186.125	115.514
	Spring	206	188.50	188.250	109.429
	Summer	92	188.00	188.750	48.742
1995	Fall	239	189.50	190.375	125.542
	Winter	221	191.25	194.125	113.844
	Spring	213	197.00		
	Summer	115			

c)

Year	Fall		Winter		Spring		Summer
1991					109.117		46.829
1992	129.387		114.128		113.577		41.931
1993	130.117		113.577		115.072		39.809
1994	130.889		115.514		109.429		48.742
1995	125.542		113.844				
Modified sum	259.504		227.973		223.006		88.760
Modified mean	129.752	+	113.987	+	111.503	+	44.380 = 399.622/400 = .99905
Seasonal index	129.875		114.095		111.609		44.422

15-32 a)

Year	Quarter	Actual Starts	4–Quarter Moving Average	Centered Moving Average	Ratio of Actual to CMA
1991	Spring	8			
	Summer	10			
	Fall	7	7.50	7.625	0.9180
	Winter	5	7.75	7.750	0.6452
			7.75		

1992	Spring	9	7.75	7.750	1.1613
	Summer	10	8.00	7.875	1.2698
	Fall	7	8.25	8.125	0.8615
	Winter	6	8.50	8.375	0.7164
1993	Spring	10	8.50	8.500	1.1765
	Summer	11	8.50	8.500	1.2941
	Fall	7	8.50	8.500	0.8235
	Winter	6	8.75	8.625	0.6957
1994	Spring	10	9.00	8.875	1.1268
	Summer	12	9.25	9.125	1.3151
	Fall	8	9.50	9.375	0.8533
	Winter	7	9.75	9.625	0.7273
1995	Spring	11	10.00	9.875	1.1139
	Summer	13	10.25	10.125	1.2840
	Fall	9			
	Winter	8			

Year	Spring		Summer		Fall		Winter	
1991					0.9180		0.6452	
1992	1.1613		1.2698		0.8615		0.7164	
1993	1.1765		1.2941		0.8235		0.6957	
1994	1.1268		1.3151		0.8533		0.7273	
1995	1.1139		1.2840					
Modified sum	2.2881		2.5781		1.7148		1.4121	
Modified mean	1.1441	+	1.2891	+	0.8574	+	0.7061	= 3.9967/4 = 0.999175
Seasonal index	1.1450		1.2902		0.8581		0.7067	

b) $\frac{.7067}{1.2902} = .5477$, so his working capital need falls by 45.23% from summer to winter.

15-34 c and d

15-36 The fact that these irregular variations even themselves out over time and the fact that they are frequently minor in magnitude enable management to live with them.

15-38 a,b)

Year	Qtr	Actual Sales	4–Quarter Moving Average	Centered Moving Average	Percentage of Sales to CMA	Seasonal Index	Deseasonalized Sales
1991	I	19				75.886	25.038
	II	24	26.50			105.081	22.840
	III	38	27.00	26.750	142.056	142.050	26.751
	IV	25	28.00	27.500	90.909	76.984	32.474
1992	I	21	29.50	28.750	73.043	75.886	27.673
	II	28	29.00	29.250	95.726	105.081	26.646
	III	44	29.50	29.250	150.427	142.050	30.975
	IV	23	30.25	29.875	76.987	76.984	29.876
1993	I	23	29.50	29.875	76.987	75.886	30.309
	II	31	29.50	29.500	105.085	105.081	29.501
	III	41	29.75	29.625	138.397	142.050	28.863
	IV	23	30.75	30.250	76.033	76.984	29.876
1994	I	24	32.50	31.625	75.889	75.886	31.626
	II	35	32.00	32.250	108.527	105.081	33.308
	III	48				142.050	33.791
	IV	21				76.984	27.278

Year	I		II		III		IV	
1991					142.056		90.909	
1992	73.043		95.726		150.427		76.987	
1993	76.987		105.085		138.397		76.033	
1994	75.889		108.527					
Modified sum	75.889	+	105.085	+	142.056	+	76.987	= 400.017/4 = 100.004
Seasonal index	75.886		105.081		142.050		76.984	

15-40 A large irregular component; a change in weather produces a larger or smaller than expected seasonal index; a change in technology which affects the secular trend; an economic change which alters the time scale of the cyclical component.

15-42 The decline in birth rates which has occurred will no doubt affect future college enrollments; we need be especially careful about the behavior in birth rates seventeen to eighteen years in the past when estimating college enrollments.

15-44 a,b)

Year	Quarter	Actual Admissions	4–Quarter Moving Average	Centered Moving Average	Ratio of Actual to CMA	Seasonal Index	Deseason-alized Admissions
1992	Spring	29				0.7851	36.9380
	Summer	30				0.8888	33.7534
			35.75				
	Fall	41		35.500	1.1549	1.1351	36.1202
			35.25				
	Winter	43		35.750	1.2028	1.1909	36.1071
			36.25				
1993	Spring	27		36.750	0.7347	0.7851	34.3905
			37.25				
	Summer	34		37.875	0.8977	0.8888	38.2538
			38.50				
	Fall	45		39.250	1.1465	1.1351	39.6441
			40.00				
	Winter	48		40.250	1.1925	1.1909	40.3057
			40.50				
1994	Spring	33		40.625	0.8123	0.7851	42.0329
			40.75				
	Summer	36		41.125	0.8754	0.8888	40.5041
			41.50				
	Fall	46		41.625	1.1051	1.1351	40.5251
			41.75				
	Winter	51		42.250	1.2071	1.1909	42.8248
			42.75				
1995	Spring	34		42.875	0.7930	0.7851	43.3066
			43.00				
	Summer	40		43.250	0.9249	0.8888	45.0045
			43.50				
	Fall	47				1.1351	41.4060
	Winter	53				1.1909	44.5042

Year	Spring		Summer		Fall		Winter	
1992					1.1549		1.2028	
1993	0.7347		0.8977		1.1465		1.1925	
1994	0.8123		0.8754		1.1051		1.2071	
1995	0.7930		0.9249					
Modified sum	0.7930	+	0.8977	+	1.1465	+	1.2028	= 4.0400/4 = 1.01
Seasonal index	0.7851		0.8888		1.1351		1.1909	

c)

Year	Quarter	Deseasonalized Admissions (Y)	x	xY	x^2
1992	Spring	36.9380	−15	−554.0695	225
	Summer	33.7534	−13	−438.7939	169
	Fall	36.1202	−11	−397.3218	121
	Winter	36.1071	−9	−324.9643	81

1993	Spring	34.3905	−7	−240.7337	49
	Summer	38.2538	−5	−191.2691	25
	Fall	39.6441	−3	−118.9323	9
	Winter	40.3057	−1	−40.3057	1
1994	Spring	42.0329	1	42.0329	1
	Summer	40.5041	3	121.5122	9
	Fall	40.5251	5	202.6253	25
	Winter	42.8248	7	299.7733	49
1995	Spring	43.3066	9	389.7593	81
	Summer	45.0045	11	495.0495	121
	Fall	41.4060	13	538.2786	169
	Winter	44.5042	15	667.5623	225
		635.6207	0	450.2030	1360

$$a = \overline{Y} = \frac{635.6207}{16} = 39.7263 \qquad b = \frac{\sum xY}{\sum x^2} = \frac{450.2030}{1360} = 0.3310$$

$\widehat{Y} = 39.7263 + 0.3310x$ (where 1990-IV 1/2 = 0 and x units = 1/2 quarter)

15-46

Year	Month	Actual Hg Level	4−Month Moving Average	Centered Moving Average
1993	Jan	0.3		
	Feb	0.7	0.650	
	March	0.8	0.750	0.7000
	April	0.8	0.750	0.7500
	May	0.7	0.700	0.7250
	June	0.7	0.650	0.6750
	July	0.6	0.575	0.6125
	Aug	0.6	0.575	0.5750
	Sept	0.4	0.475	0.5250
	Oct	0.7	0.450	0.4625
	Nov	0.2	0.450	0.4500
	Dec	0.5	0.500	0.4750
1994	Jan	0.4	0.625	0.5625
	Feb	0.9	0.725	0.6750
	March	0.7	0.750	0.7375
	April	0.9	0.725	0.7375
	May	0.5	0.725	0.7250
	June	0.8	0.675	0.7000

Year	Month	Actual Hg Level	4−Month Moving Average	Centered Moving Average
1994	July	0.7	0.650	0.6625
	Aug	0.7	0.600	0.6250
	Sept	0.4	0.500	0.5500
	Oct	0.6	0.425	0.4625
	Nov	0.3	0.375	0.4000
	Dec	0.4	0.375	0.3750
1995	Jan	0.2	0.450	0.4125
	Feb	0.6	0.575	0.5125
	March	0.6	0.700	0.6375
	April	0.9	0.725	0.7125
	May	0.7	0.775	0.7500
	June	0.7	0.750	0.7625
	July	0.8	0.700	0.7250
	Aug	0.8	0.675	0.6875
	Sept	0.5	0.550	0.6125
	Oct	0.6	0.475	0.5125
	Nov	0.3		
	Dec	0.5		

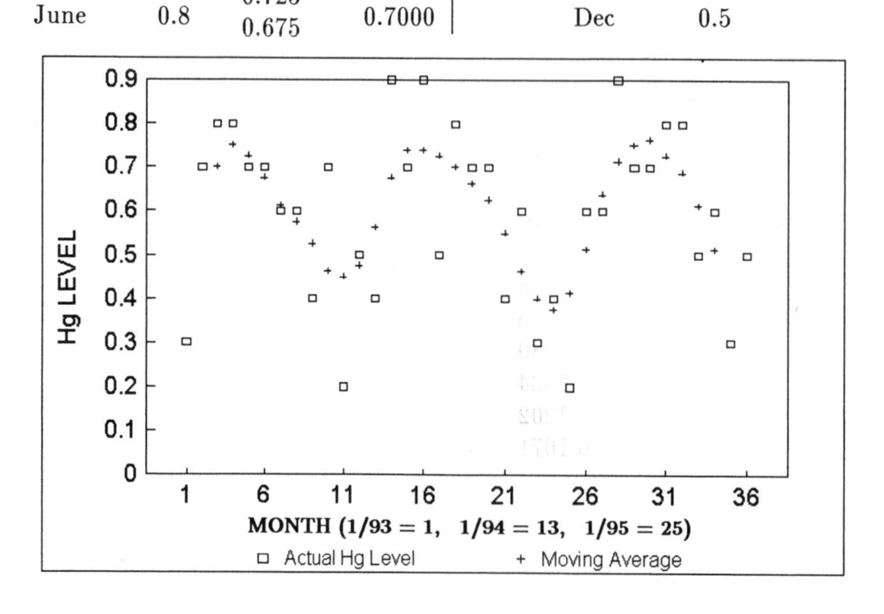

15-48 a) Gasoline mileage is affected by such things as government responses to the 1973 oil embargo and the resultant mandated fleet mileage standards.

b) This series is almost entirely irregular variation, because commercial aviation fatalities occur in random batches as the result of unpredictable airplane crashes.

c) Although total world demand has a long-run increasing trend, there are so many grain growers that each one's exports does not grow smoothly over time but depends instead on political and economic conditions in both importing and exporting nations.

d) In addition to seasonalities resulting from higher usages in the summer months, gasoline prices are also greatly affected by unpredictable geopolitical events.

15-50 a) Although sales of PC's have been growing at increasing rates, this growth cannot be sustained as even larger fractions of the population eventually come to own PC's. Because of this, a second-degree predicting equation will soon tend to overestimate the sales of PC's.

b) Here, too, a forecast based on a second-degree predicting equation will tend to be an overestimate, because of the saturation phenomenon mentioned in (a) and also because kids will tend to play with them less as the novelty of the games wears off.

c) As more states act to place caps on giving awards for damages in medical malpractice cases, the amounts paid for such claims will cease its current rapid growth. As insurance companies' liabilities stop growing so rapidly, so will the premiums they charge. Once again, second-degree forecasts will tend to be overestimated as a result.

d) Here is another instance of a growth rate that cannot be sustained which will lead to overestimates from a second-degree predicting equation.

15-52 a)

Year	Qtr	Actual Exports	4–Quarter Moving Average	Centered Moving Average	Percentage of Actual to CMA	Seasonal Index	Deseason-alized Exports
1992	I	1				43.343	2.307
	II	3	3.50			68.730	4.365
	III	6	3.75	3.625	165.517	173.374	3.461
	IV	4	3.50	3.625	110.345	114.551	3.492
1993	I	2	3.75	3.625	55.172	43.343	4.614
	II	2	4.00	3.875	51.613	68.730	2.910
	III	7	4.00	4.000	175.000	173.374	4.037
	IV	5	4.50	4.250	117.647	114.551	4.365
1994	I	2	4.75	4.625	43.243	43.343	4.614
	II	4	4.75	4.750	84.211	68.730	5.820
	III	8	4.50	4.625	172.973	173.374	4.614
	IV	5	4.25	4.375	114.286	114.551	4.365
1995	I	1	4.25	4.250	23.529	43.343	2.307
	II	3	4.50	4.375	68.571	68.730	4.365
	III	8				173.374	4.614
	IV	6				114.551	5.238

Year	I	II	III	IV	
1992			165.517	110.345	
1993	55.172	51.613	175.000	117.647	
1994	43.243	84.211	172.973	114.286	
1995	23.529	68.571			
Modified sum	43.243	+ 68.571	+ 172.973	+ 114.286	= 399.073/4 = 99.76825
Seasonal index	43.343	68.730	173.375	114.551	

b,c)

Year	Qtr	Deseasonalized Exports(Y)	x	xY	x^2	Deseasonalized trend $\widehat{Y}$=4.0930 +0.0433x	Relative Cyclical Residual $100(Y/\widehat{Y}-1)$
1992	I	2.307	−15	−34.605	225	3.444	−33.014
	II	4.365	−13	−56.745	169	3.530	23.654
	III	3.461	−11	−38.071	121	3.617	−4.313
	IV	3.492	−9	−31.428	81	3.703	−5.698
1993	I	4.614	−7	−32.298	49	3.790	21.741
	II	2.910	−5	−14.550	25	3.877	−24.942
	III	4.037	−3	−12.111	9	3.963	1.867
	IV	4.365	−1	−4.365	1	4.050	7.778
1994	I	4.614	1	4.614	1	4.136	11.557
	II	5.820	3	17.460	9	4.223	37.817
	III	4.614	5	23.070	25	4.310	7.053
	IV	4.365	7	30.555	49	4.396	−0.705
1995	I	2.307	9	20.763	81	4.483	−48.539
	II	4.365	11	48.015	121	4.569	−4.465
	III	4.614	13	59.982	169	4.656	−0.902
	IV	5.238	15	78.570	225	4.743	10.436
		65.488	0	58.856	1360		

$a = \overline{Y} = \frac{65.488}{16} = 4.0930$ $\qquad b = \frac{\sum xY}{\sum x^2} = \frac{58.856}{1360} = .0433$

$\widehat{Y} = 4.0930 + .0433x$ (where 1993-IV 1/2 = 0 and x units = 1/2 quarter)

d)

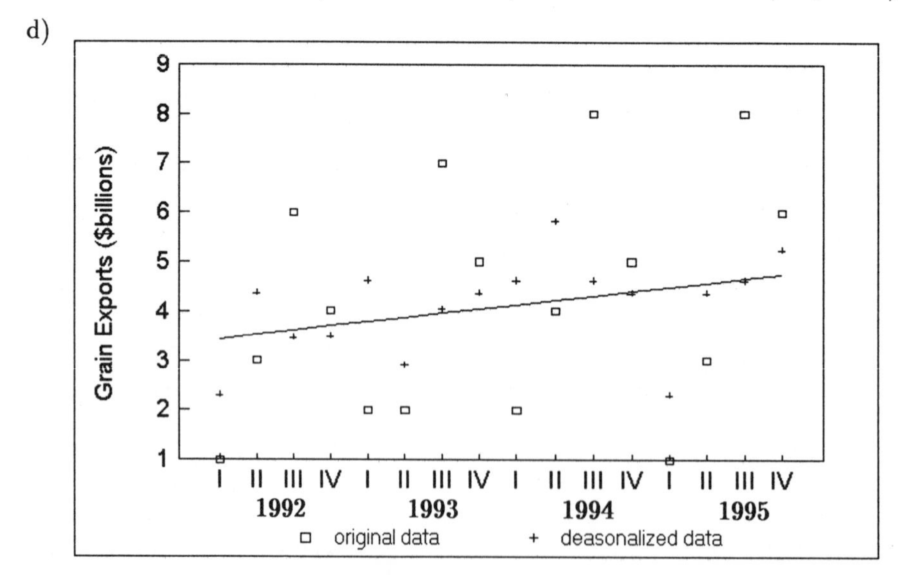

15-54 Since such a major source of demand for heavy earth-moving equipment is going to be lost, historic trends in sales of such equipment will be poor predictors of future sales. The manufacturers would be better advised to abandon a times series forecasting model for an econometric model which includes such explanatory variables as the number of miles of roads currently under construction and scheduled for the next few years, the age of the current stocks of earth-moving equipment, etc.

15-56 a)

Year	x	Y	xY	x^2
1988	−7	12	−84	49
1989	−5	11	−55	25
1990	−3	19	−57	9
1991	−1	17	−17	1
1992	1	19	19	1
1993	3	18	54	9
1994	5	20	100	25
1995	7	23	161	49
	0	139	121	168

$a = \overline{Y} = \frac{139}{8} = 17.3750$ $\qquad b = \frac{\sum xY}{\sum x^2} = \frac{121}{168} = 0.7202$

$\widehat{Y} = 17.3750 + 0.7202x$ (where 1991.5 = 0 and x units = .5 year)

b) In 1999 $\widehat{Y} = 17.3750 + 0.7202(15) = 28.2$, or about 28 completions.

c) He should be very careful about predicting so far in advance because of the many things that can change in the home-building business in the meantime.

15-58 a,b)

Year	Qtr	Deseasonalized Unemployment(Y)	x	xY	x^2	Deseasonalized trend $\widehat{Y}$=8.175 −0.0379x	Percent of trend 100($Y/\widehat{Y}$)
1991	I	7.3	−19	−138.7	361	8.8951	82.1
	II	7.2	−17	−122.4	289	8.8193	81.6
	III	7.3	−15	−109.5	225	8.7435	83.5
	IV	8.1	−13	−105.3	169	8.6677	93.5
1992	I	8.7	−11	−95.7	121	8.5919	101.3
	II	9.2	−9	−82.8	81	8.5161	108.0
	III	9.8	−7	−68.6	49	8.4403	116.1
	IV	10.5	−5	−52.5	25	8.3645	125.5
1993	I	10.2	−3	−30.6	9	8.2887	123.1
	II	9.9	−1	−9.9	1	8.2129	120.5
	III	9.2	1	9.2	1	8.1371	113.1
	IV	8.3	3	24.9	9	8.0613	103.0
1994	I	7.6	5	38.0	25	7.9855	95.2
	II	7.4	7	51.8	49	7.9097	93.6
	III	7.5	9	67.5	81	7.8339	95.7
	IV	7.6	11	83.6	121	7.7581	98.0
1995	I	7.4	13	96.2	169	7.6823	96.3
	II	7.0	15	105.0	225	7.6065	92.0
	III	6.8	17	115.6	289	7.5307	90.3
	IV	6.5	19	123.5	361	7.4549	87.2
		163.5	0	−100.7	2660		

$a = \overline{Y} = \frac{163.5}{20} = 8.175$ $\qquad b = \frac{\sum xY}{\sum x^2} = \frac{-100.7}{2660} = -0.0379$

$\widehat{Y} = 8.175 - 0.0379x$ (where 1993 - II 1/2 = 0 and x units = 1/2 quarter)

c)

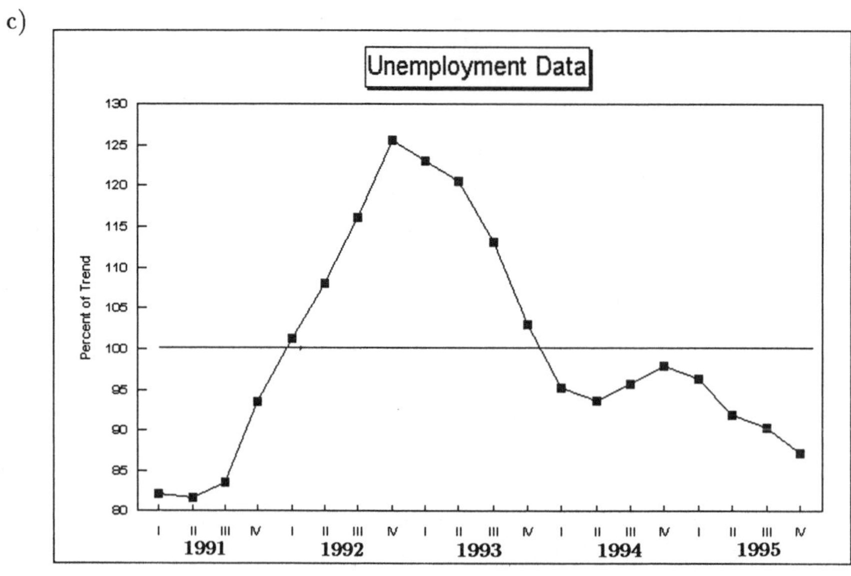

15-60 a)

Week	Sales (Y)	x	xY	x^2
1	41	−3	−123	9
2	52	−2	−104	4
3	79	−1	−79	1
4	76	0	0	0
5	72	1	72	1
6	59	2	118	4
7	41	3	123	9
	420	0	7	28

$a = \overline{Y} = \frac{420}{7} = 60$ $\qquad b = \frac{\sum xY}{\sum x^2} = \frac{7}{28} = 0.25$

$\widehat{Y} = 60 + 0.25x$

b) For week 8, $\widehat{Y} = 60 + 0.25(4) = 61$

c) The data suggest that sales of the broiled whole chickens have peaked and are now declining, so a 2nd-degree curve would have have been better. The linear regression line predicts sales will continue to increase.

15-62 a) Multiplying through by (SEASONAL INDICES)/100, we get the actual passenger counts:

	Spring	Summer	Fall	Winter
1994	652.30	397.85	689.30	598.00
1995	704.00	408.80	678.00	577.20

b) Summer saw the fewest passengers; spring saw the most.

c) For the 1996 fall season, $x = 13$ and

$\widehat{Y} = 584.75 - .45(13) = 578.9$ (deseasonalized)

so the actual predicted ridership is 578.9(113/100) = 654.157, or about 654,000 riders.

15-64

Week	Day	Actual Customers	Moving Average	Percentage of Actual to MA
1	MON	345		
	TUE	310		
	WED	385		
	THU	416	487.4286	85.3458
	FRI	597	497.8571	119.9139
	SAT	706	501.1429	140.8780
	SUN	653	503.2857	129.7474
2	MON	418	517.4286	80.7841
	TUE	333	527.0000	63.1879
	WED	400	534.8571	74.7863
	THU	515	541.8571	95.0435
	FRI	664	538.2857	123.3546
	SAT	761	546.0000	139.3773
	SUN	702	533.2857	131.6368
3	MON	393	536.1429	73.3014
	TUE	387	530.5714	72.9402
	WED	311	523.4286	59.4159
	THU	535	508.5714	105.1966
	FRI	625	510.4286	122.4461
	SAT	711	514.0000	138.3268
	SUN	598	523.4286	114.2467
4	MON	406	510.4286	79.5410
	TUE	412	514.0000	80.1556
	WED	377	527.1429	71.5176
	THU	444	559.1429	79.4072
	FRI	650		
	SAT	803		
	SUN	822		

Week	MON	TUE	WED	THU	FRI	SAT	SUN
1				85.3458	119.9139	140.8780	129.7474
2	80.7841	63.1879	74.7863	95.0435	123.3546	139.3773	131.6368
3	73.3014	72.9402	59.4159	105.1966	122.4461	138.3268	114.2467
4	79.5410	80.1556	71.5176	79.4072			
Modified sum	79.5410	72.9402	71.5176	180.3893	122.4461	139.3773	129.7474
Modified mean	79.5410 +	72.9402 +	71.5176 +	90.1947 +	122.4461 +	139.3773 +	129.7474
Seasonal index	78.8914	72.3445	70.9335	89.4580	121.4460	138.2389	128.6877

(Sum of modified means = 705.7643/7 = 100.82347)

15-66

Year	1991	1992	1993	1994	1995
x	−2	−1	0	1	2
Actual sales (Y)	3.5	3.8	4.0	3.7	3.9
Trend($\widehat{Y}$=3.78+.07x)	3.64	3.71	3.78	3.85	3.92
% of trend ($100Y/\widehat{Y}$)	96.15	102.43	105.82	96.10	99.49

1993 had the largest percent of trend, 1994 the smallest.

CHAPTER 16

INDEX NUMBERS

16-2 Price indices and quantity indices describe the change in a single variable, price and quantity (or number), respectively. A value index describes how the product of the two variables, price and quantity, changes over a period of time.

16-4 An index may be used by itself or as a part of an intermediate computation to better understand some other information.

16-6 $$\text{Percentage relative} = \frac{\text{Current Value}}{\text{Base Value}} \times 100$$

16-8

	1992 P_0	1993 P_1	1994 P_2	1995 P_3
Class A	8.48	9.32	10.34	11.16
Class B	6.90	7.52	8.19	8.76
Class C	4.50	4.99	5.48	5.86
Class D	3.10	3.47	3.85	4.11
	22.98	25.30	27.86	29.89
Index $= \frac{\sum P_i}{\sum P_0} \times 100$:	$\frac{2298}{22.98}$ = 100.0	$\frac{2530}{22.98}$ = 110.1	$\frac{2786}{22.98}$ = 121.2	$\frac{2989}{22.98}$ = 130.1

16-10

Commodity	1993 P_1	1994 P_0	1995 P_2
Dairy	$2.34	$2.38	$2.60
Meat	3.19	3.41	3.36
Vegetables	.85	.89	.94
Fruit	1.11	1.19	1.18
	$7.49	$7.87	$8.08
Index $= \frac{\sum P_i}{\sum P_0} \times 100$:	$\frac{749}{7.87}$ = 95.2	$\frac{787}{7.87}$ = 100.0	$\frac{808}{7.87}$ = 102.7

16-12

Transaction	Q_0 1994	Q_1 1995
Savings withdrawal	169,000	158,000
Checking withdrawal	21,843,000	23,241,000
Savings deposit	293,000	303,000
Checking deposit	2,684,000	3,361,000
	24,989,000	27,063,000

$$\text{1995 Index} = \frac{\sum Q_1}{\sum Q_0} \times 100 = \frac{2{,}706{,}300{,}000}{24{,}989{,}000} = 108.3$$

Data for Problems 16-14 to 16-16

	Selling Price				Number Sold			
	1993	1994	1995	1996	1993	1994	1995	1996
ED 107	1894	1906	1938	1957	84.6	86.9	96.4	107.5
ED Electra	2506	2560	2609	2680	38.4	42.5	55.6	67.5
ED Optima	1403	1440	1462	1499	87.4	99.4	109.7	134.6
ED 821	1639	1650	1674	1694	75.8	78.9	82.4	86.4

16-14 Laspeyres index

Base year: 1993	1993 P_0Q_0	1994 P_1Q_0	1995 P_2Q_0	1996 P_3Q_0
ED 107	160,232.4	161,247.6	163,954.8	165,562.2
ED Electra	96,230.4	98,304.0	100,185.6	102,912.0
ED Optima	122,622.2	125,856.0	127,778.8	131,012.6
ED 821	124,236.2	125,070.0	126,889.2	128,405.2
	503,321.2	510,477.6	518,808.4	527,892.0
Index $= \frac{\sum P_iQ_0}{\sum P_0Q_0} \times 100$:	$\frac{50,332,120}{503,321.2}$ = 100.0	$\frac{51,047,760}{503,321.2}$ = 101.4	$\frac{51,880,840}{503,321.2}$ = 103.1	$\frac{52,789,200}{503,321.2}$ = 104.9

16-16 Paasche index

Base period: 1994	1993 P_1Q_1	1993 P_0Q_1	1995 P_2Q_2	1995 P_0Q_2	1996 P_3Q_3	1996 P_0Q_3
ED 107	160,232.4	161,247.6	190,699.2	187,550.4	210,377.5	204,895.0
ED Electra	96,230.4	98,304.0	145,060.4	142,336.0	180,900.0	172,800.0
ED Optima	122,622.2	125,856.0	160,381.4	157,968.0	201,765.4	193,824.0
ED 821	124,236.2	125,070.0	137,937.6	135,960.0	146,361.6	142,560.0
	503,321.2	510,477.6	634,078.6	623,814.4	739,404.5	714,079.0

$$\text{Index} = \frac{\sum P_iQ_i}{\sum P_0Q_i} \times 100: \quad \frac{50,332,120}{510,477.6} = 98.6 \qquad \frac{63,407,860}{623,814.4} = 101.6 \qquad \frac{73,940,450}{714,079.0} = 103.5$$

16-18

Fruit	June P_0	July P_1	August P_2	June Q_0	June P_0Q_0	July P_1Q_0	August P_2Q_0
Apples	.59	.64	.69	150	88.50	96.00	103.50
Oranges	.75	.65	.70	200	150.00	130.00	140.00
Peaches	.87	.90	.85	125	108.75	112.50	106.25
Watermelons	1.00	1.10	.95	350	350.00	385.00	332.50
Canteloupes	.95	.89	.90	150	142.50	133.50	135.00
					839.75	857.00	817.25
Index $= \frac{\sum P_iQ_0}{\sum P_0Q_0} \times 100$:					$\frac{83975}{839.75}$ = 100.00	$\frac{85700}{839.75}$ = 102.1	$\frac{81725}{839.75}$ = 97.30

They are Laspeyres indices, since the weights are the base-month quantities.

16-20

Type of seats	1992 P_1	1993 P_0	1994 P_2	1995 P_3	1993 Q_0
Box seats	6.50	7.25	7.50	8.10	27
General admission	3.50	3.85	4.30	4.35	80

$$\text{Index} = \frac{\sum\left(\frac{P_i}{P_0} \times 100\right)(P_0Q_0)}{\sum P_0Q_0} = \left(\frac{\sum P_iQ_0}{\sum P_0Q_0}\right) \times 100$$

P_1Q_0	P_0Q_0	P_2Q_0	P_3Q_0
175.50	195.75	202.50	218.70
280.00	308.00	344.00	348.00
455.50	503.75	546.50	566.70

Index: $\frac{45550}{503.75} = 90.4$ $\quad \frac{50375}{503.75} = 100.0$ $\quad \frac{54650}{503.75} = 108.5$ $\quad \frac{56670}{503.75} = 112.5$

16-22

Repair	1991 P_0	1993 P_1	1995 P_2	1991 P_0/P_0	1993 P_1/P_0	1995 P_2/P_0
Water pump	35	37	41	1.000	1.057	1.171
Engine valves	189	205	216	1.000	1.085	1.143
Wheel balancing	26	29	30	1.000	1.115	1.154
Tune-up	16	16	18	1.000	1.000	1.125
				4.000	4.257	4.593

$$\text{Index} = \frac{\sum\left(\frac{P_i}{P_0} \times 100\right)}{n}: \quad \frac{400.0}{4} = 100.0 \quad \frac{425.7}{4} = 106.4 \quad \frac{459.3}{4} = 114.8$$

16-24

Product	1994 P_0	1996 P_1	$\frac{P_1}{P_0}$	P_0Q_0	$\left(\frac{P_1}{P_0}\right)(P_0Q_0)$
1-megabyte chips	\$ 42	\$ 65	1.5476	957	1481.07
4-megabyte chips	180	247	1.3722	487	668.27
16-megabyte chips	447	612	1.3691	349	477.83
			4.2889	1793	2627.17

a) $\text{Index} = \frac{\sum\left(\frac{P_i}{P_0} \times 100\right)}{n} = \frac{428.89}{3} = 143.0$

b) $\text{Index} = \frac{\sum\left(\frac{P_i}{P_0} \times 100\right)(P_0Q_0)}{\sum P_0Q_0} = \frac{262,717}{1793} = 146.5$

16-26

Group	1992 P_1	1993 P_2	1994 P_0	1995 P_3	1992 P_1/P_0	1993 P_2/P_0	1994 P_0/P_0	1995 P_3/P_0
Physicians	54	65	86	103	0.628	0.756	1.000	1.198
Student	39	41	55	76	0.709	0.745	1.000	1.382
Government	48	61	76	93	0.632	0.803	1.000	1.224
Teachers	46	58	75	96	0.613	0.773	1.000	1.280
					2.582	3.077	4.000	5.084

$$\text{Index} = \frac{\sum\left(\frac{P_i}{P_0} \times 100\right)}{n}: \quad \frac{258.2}{4} = 64.5 \quad \frac{307.7}{4} = 76.9 \quad \frac{400.0}{4} = 100.0 \quad \frac{508.4}{4} = 127.1$$

16-28

Maps	1993 P_0	1994 P_1	1995 P_2	1993 Q_0	1993 P_0Q_0	1994 P_1Q_0	1995 P_2Q_0
City	.75	.90	1.10	1000	750	900	1100
County	.75	.90	1.00	400	300	360	400
State	1.00	1.50	1.50	1000	1000	1500	1500
U.S.	2.50	2.75	2.75	220	550	605	605
					2600	3365	3605

$$\text{Index} = \frac{\sum\left(\frac{P_i}{P_0} \times 100\right)(P_0Q_0)}{\sum P_0Q_0} = \frac{\sum P_iQ_0}{\sum P_0Q_0} \times 100: \quad \frac{336,500}{2600} = 129.4 \quad \frac{360,500}{2600} \doteq 138.7$$

16-30 The weighted aggregates index uses quantities for weights whereas the weighted average of relatives index uses values for weights.

16-32

	1993	1994	1995	1995	1993	1994	1995
Product	Q_1	Q_2	Q_0	P_0	Q_1P_0	Q_2P_0	Q_0P_0
Barges	92	118	85	33	3036	3894	2805
Cars	456	475	480	56	25536	26600	26880
Trucks	52	56	59	116	6032	6496	6844
					34604	36990	36529

$$\text{Index} = \frac{\sum Q_iP_0}{\sum Q_0P_0} \times 100: \quad \frac{3{,}460{,}400}{36529} = 94.7 \quad \frac{3{,}699{,}000}{36529} = 101.3 \quad \frac{3{,}652{,}900}{36529} = 100.0$$

16-34

	1992	1993	1994	1995	1992	1993	1994	1995
Type of Crime	Q_1	Q_2	Q_3	Q_0	Q_1/Q_0	Q_2/Q_0	Q_3/Q_0	Q_0/Q_0
Assault & Rape	110	128	134	129	0.853	0.992	1.039	1.000
Murder	30	45	40	48	0.625	0.938	0.833	1.000
Robbery	610	720	770	830	0.735	0.867	0.928	1.000
Larceny	2450	2630	2910	2890	0.848	0.910	1.007	1.000
					3.061	3.707	3.808	4.000

$$\text{Index} = \frac{\sum\left(\frac{Q_i}{Q_0} \times 100\right)}{n}: \quad \frac{306.1}{4} = 76.5 \quad \frac{370.7}{4} = 92.7 \quad \frac{380.8}{4} = 95.2 \quad \frac{400.0}{4} = 100.0$$

16-36

	1973	1973	1993	1973	1993
Age (yrs.)	"P_0"	Q_0	Q_1	P_0Q_0	P_0Q_1
< 4	5000	400	125	2,000,000	625,000
4 – 15	4000	295	200	1,180,000	800,000
16 – 25	24000	1230	1000	29,520,000	24,000,000
36 – 60	19000	700	450	13,300,000	8,550,000
> 60	7000	1100	935	7,700,000	6,545,000
				53,700,000	40,520,000

$$\text{1993 index} = \frac{\sum P_0Q_1}{\sum P_0Q_0} \times 100: \quad \frac{4{,}052{,}000{,}000}{53{,}700{,}000} = 75.5$$

16-38 Appropriate weighting for one period may become inappropriate in a short time. Unless the weights are changed the index becomes less informative.

16-40 The values from several adjoining periods are averaged.

16-42 An index does not reflect changes in the quality of items and therefore understates or overstates the price level change if the quality changes.

16-44

	1991	1993	1995
Commodity	V_0	V_1	V_2
Coffee	834	1436	1321
Sugar	96	118	122
Copper	241	258	269
Zinc	142	125	106
	1313	1937	1818

$$\text{Index} = \frac{\sum V_i}{\sum V_0} \times 100: \quad \frac{131300}{1313} = 100.0 \quad \frac{193700}{1313} = 147.5 \quad \frac{181800}{1313} = 138.5$$

16-46

Product	1991 P_0	1995 P_1	1991 Q_0	1991 P_0Q_0	1995 P_1Q_0
Cheese	1.45	1.49	2.6	3.77	3.874
Milk	1.60	1.61	47.6	76.16	76.636
Butter	.70	.80	3.1	2.17	2.480
				82.10	82.990

$$\text{1995 index} = \frac{\sum P_1Q_0}{\sum P_0Q_0} \times 100 = \frac{8299}{82.10} = 101.1$$

16-48 The problem of incomparability of indices would be present; there has been a basic change in what is being measured by the indices, because computer technology has changed significantly over the past few decades.

16-50

Model	1993 Q_0	1994 Q_1	1995 Q_2	1993 P_0	1993 Q_0P_0	1994 Q_1P_0	1995 Q_0P_0
Sport	45	48	56	89	4005	4272	4984
Touring	64	67	71	104	6656	6968	7384
Cross Country	28	35	27	138	3864	4830	3726
Sprint	21	16	28	245	5145	3920	6860
					19670	19990	22954

$$\text{Index} = \frac{\sum\left(\frac{Q_i}{Q_0} \times 100\right)(Q_0P_0)}{\sum Q_0P_0} = \left(\frac{\sum Q_iP_0}{\sum Q_0P_0}\right) \times 100:$$

$\frac{1{,}967{,}000}{19670} = 100.0$ $\quad \frac{1{,}999{,}000}{19670} = 101.6$ $\quad \frac{2{,}295{,}400}{19670} = 116.7$

16-52

Costs	1991 V_1	1993 V_0	1995 V_2
Wages	24,378	36,421	37,613
Lumber	1,816	2,019	2,136
Utilities	638	681	701
	26,832	39,121	40,450

$$\text{Index} = \frac{\sum V_i}{\sum V_0} \times 100: \quad \frac{2{,}683{,}200}{39121} = 68.6 \quad \frac{3{,}912{,}100}{39121} = 100.0 \quad \frac{4{,}045{,}000}{39121} = 103.4$$

16-54 Depending on what is being measured, the choice of base periods can significantly distort the importance of a particular value.

16-56

Product	1991 P_0	1991 Q_0	1992 Q_1	1993 Q_2	1994 Q_3	1995 Q_4
Wheat	4.40	610	620	640	630	650
Corn	3.60	390	390	410	440	440
Oats	1.20	100	90	120	130	150
Rye	24.00	10	20	10	10	20
Barley	2.10	160	150	120	190	180
Soybeans	5.60	130	140	160	120	130

1991	1992	1993	1994	1995
Q_0P_0	Q_1P_0	Q_2P_0	Q_3P_0	Q_4P_0
2684	2728	2816	2772	2860
1404	1404	1476	1584	1584
120	108	144	156	180
240	480	240	240	480
336	315	252	399	378
728	784	896	672	728
5512	5819	5824	5823	6210

Index $= \frac{\sum Q_i P_0}{\sum Q_0 P_0} \times 100$: $\frac{551200}{5512} = 100.0$, $\frac{581900}{5512} = 105.6$, $\frac{582400}{5512} = 105.7$, $\frac{582300}{5512} = 105.6$, $\frac{621000}{5512} = 112.7$

16-58

Size	1991 P_1	1991 Q_1	1993 P_0	1993 Q_0	1995 P_2	1995 Q_2
Subcompact	62	32	68	65	70	86
Compact	76	45	78	68	80	73
Sedan	90	462	98	325	106	386

a)

1991	1993	1995
P_1Q_0	P_0Q_0	P_2Q_0
4030	4420	4550
5168	5304	5440
29250	31850	34450
38448	41574	44440

Index $= \frac{\sum\left(\frac{P_i}{P_0} \times 100\right)(P_0Q_0)}{\sum P_0Q_0} = \frac{\sum P_iQ_0}{\sum P_0Q_0} \times 100$: $\frac{3,844,800}{41574} = 92.5$, $\frac{4,157,400}{41574} = 100.0$, $\frac{4,444,000}{41574} = 106.9$

b)

1991 P_1Q_1	1991 $\left(\frac{P_1}{P_0}\right)P_1Q_1$	1993 $\left(\frac{P_0}{P_0}\right)P_0Q_0$	1995 P_2Q_2	1995 $\left(\frac{P_2}{P_0}\right)P_2Q_2$
1984	1808.94	4420	6020	6197.06
3420	3332.30	5304	5840	5989.74
41580	38185.72	31850	40916	44256.08
46984	43326.96	41574	52776	56442.88

Index $= \frac{\sum\left(\frac{P_i}{P_0} \times 100\right)(P_iQ_i)}{\sum P_iQ_i}$: $\frac{4,332,696}{46984} = 92.2$, $\frac{4,157,400}{41574} = 100.0$, $\frac{5,644,288}{52776} = 106.9$

16-60

Department	1994 P_0	1995 P_1	1996 P_2	1994 P_0/P_0	1995 P_1/P_0	1996 P_2/P_0
Mechanical	3642	3891	4253	1.000	1.068	1.168
Chemical	3888	4052	4425	1.000	1.042	1.138
Biomedical	4251	4537	4724	1.000	1.067	1.111
Electrical	3764	4305	4297	1.000	1.144	1.142
				4.000	4.321	4.559

Index $= \frac{\sum\left(\frac{P_i}{P_0} \times 100\right)}{n}$: $\frac{400.0}{4} = 100.0$, $\frac{432.1}{4} = 108.0$, $\frac{455.9}{4} = 114.0$

16-62 1993 real average wage = 521.35(100)/152 = $342.99

16-64

Subject	1993 Q_1	1994 Q_2	1995 Q_3	1996 Q_0	1993 Q_1/Q_0	1994 Q_2/Q_0	1995 Q_3/Q_0	1996 Q_0/Q_0
Local	73	76	112	107	0.682	0.710	1.047	1.000
Snowboard	101	129	163	162	0.623	0.796	1.006	1.000
Handicapped	163	189	271	268	0.608	0.705	1.011	1.000
Regular	183	210	303	298	0.614	0.705	1.017	1.000
					2.527	2.916	4.081	4.000

$$\text{Index} = \frac{\sum\left(\frac{Q_i}{Q_0} \times 100\right)}{n}: \quad \frac{252.7}{4} = 63.2 \quad \frac{291.6}{4} = 72.9 \quad \frac{408.1}{4} = 102.0 \quad \frac{400.0}{4} = 100.0$$

16-65

Product	1992 Q_1	1993 Q_2	1994 Q_0	1995 Q_3	1994 P_0	1992 Q_1P_0	1993 Q_2P_0	1994 Q_3P_0	1995 Q_0P_0
Wheat	4.6	6.7	4.0	5.2	2680	12328	17956	10720	13936
Feed grains	4.9	6.2	1.8	1.2	2270	11123	14074	4086	2724
Soybeans	4.7	5.7	1.2	1.8	3430	16121	19551	4116	6174
						39572	51581	18922	22834

$$\text{Index} = \frac{\sum Q_iP_0}{\sum Q_0P_0} \times 100: \quad \frac{3957200}{18922} = 209.1 \quad = \frac{5158100}{18922} = 272.6 \quad \frac{1892200}{18922} = 100.0 \quad \frac{2283400}{18922} = 120.7$$

16-66

Time	1991 P_0	1991 Q_0	1996 P_1	P_1Q_1	P_0Q_1
Day	.17	5.2	.19	0.884	0.988
Evening	.13	8.7	.16	1.131	1.392
Night	.09	10.3	.12	0.927	1.236
				2.942	3.616

$$\text{1996 Laspeyres index} = \frac{\sum P_1Q_0}{\sum P_0Q_0} \times 100 = \frac{361.6}{2.942} = 122.9$$

16-67

Town	1992 P_1	1994 P_0	1996 P_2	1994 Q_0	1992 P_1Q_0	1994 P_0Q_0	1996 P_2Q_0
Greenville	21206	24210	26235	17	360502	411570	445995
Hampton	17129	19722	22109	14	239806	276108	309526
Middletown	25723	28657	32481	21	540183	601797	682101
					1140491	1289475	1437622

$$\text{Index} = \frac{\sum P_iQ_0}{\sum P_0Q_0} \times 100: \quad \frac{114049100}{1289475} = 88.4 \quad \frac{128947500}{1289475} = 100.0 \quad \frac{143762200}{1289475} = 111.5$$

16-68

Item	1993 P_0	1994 P_1	1995 P_2	1996 P_3	1996 Q	1993 P_0Q	1994 P_1Q	1995 P_2Q	1996 P_3Q
Hamburger	0.58	0.62	0.69	0.79	1.8	1.044	1.116	1.242	1.422
Chicken	1.89	2.09	2.18	2.25	2.1	3.969	4.389	4.578	4.725
French fries	0.84	0.89	0.99	0.99	2.4	2.016	2.136	2.376	2.376
Onion rings	0.91	0.99	1.14	1.19	1.6	1.456	1.584	1.824	1.904
						8.485	9.225	10.020	10.427

$$\text{Index} = \frac{\sum P_iQ}{\sum P_0Q} \times 100: \quad \frac{848.5}{8.485} = 100.0 \quad \frac{922.5}{8.485} = 108.7 \quad \frac{1002.0}{8.485} = 118.1 \quad \frac{1042.7}{8.485} = 122.9$$

16-70 Doubling a factor weight gives that factor extra impact in lieu of the missing factor; assigning low scores to a missing factor calls into question the entire rating process. Alternative responses to missing data include leaving out schools with missing information, or assigning average values to the missing factors. However, these alternatives still produce some distortions in the ratings.

CHAPTER 17

DECISION THEORY

17-2 Lisa is correct if she can obtain certain information; otherwise she isn't. She must determine her objective (presumably to maximize Adventures, Inc.'s profit), the available courses of action (which investments to make), the payoffs from these actions and the probability of the various payoffs being realized. The last two of these most likely will be difficult to determine.

17-4 We assume that the mechanics are paid for their vacations. Note that each mechanic works 2000 hours/year (50 weeks @ 40 hours/week).

a) The payoff table below gives both conditional and expected profits.

Mechanics needed		5	6	7	8	Expected
Probability		.2	.3	.4	.1	profit
	5	66400	66400	66400	66400	66400
Mechanics	6	47680	79680	79680	79680	73280 ←
hired	7	28960	60960	92960	92960	70560
	8	10240	42240	74240	106240	55040

So the optimal decision is to hire six mechanics.

b) EVPI = .2(66400) + .3(79680) + .4(92960) + .1(106240) − 73280 = $11,712

17-6 Labor cost = .375 × 72 boxes = $27 per case sold
Purchase cost = 60 per case
$87 total cost if sold

1.50 × 72 = $108 = Total revenue per case sold
87 = Total cost
$ 21 = Total profit per case

The payoff table below gives both conditional and expected profits.

Cases sold		15	16	17	18	19	20	Expected
Probability		.05	.20	.30	.25	.10	.10	profit
	15	315	315	315	315	315	315	315.00
	16	255	336	336	336	336	336	331.95
Cases	17	195	276	357	357	357	357	332.70 ←
Stocked	18	135	216	297	378	378	378	309.15
	19	75	156	237	318	399	399	265.35
	20	15	96	177	258	339	420	213.45

a) Order 17 cases

b) Expected profit $332.70

17-8 The payoff table below gives both conditional and expected profits.

Demand (dozens)		55	56	57	58	59	60	Expected
Probability		0.15	0.20	0.10	0.35	0.15	0.05	profit
	55	646.80	646.80	646.80	646.80	646.80	646.80	646.80
	56	637.56	658.56	658.56	658.56	658.56	658.56	655.41
Pizzas	57	628.32	649.32	670.32	670.32	670.32	670.32	659.82
Ordered	58	619.08	640.08	661.08	682.08	682.08	682.08	662.13 ←
(dozens)	59	609.84	630.84	651.84	672.84	693.84	693.84	657.09
	60	600.60	621.60	642.60	663.60	684.60	705.60	648.90

They should order 58 dozen pizzas.

$$\text{EVPI} = 646.80(.15) + 658.56(.20) + 670.32(.10) + 682.08(.35) + 693.84(.15) + 705.60(.05) - 662.13 = \$11.718$$

Thus \$11.72 is the maximum they should pay for perfect information about the demand.

17-10 D.O.T. is \$8 better off on its budget for every sign it buys at \$21. This is similar to a marginal profit. \$21 is the cost D.O.T. has to absorb if it overstocks. This is the marginal loss.

$p^* = \dfrac{\text{ML}}{\text{MP} + \text{ML}} = \dfrac{21}{29} = .7241,$ which corresponds to $-.595\sigma$, so D.O.T.should purchase $\mu - .595\sigma = 78 - .595(15) = 69.075$, or 69 signs.

17-12 a) $\text{MP} = 1.50 - 0.67 = 0.83$ $\quad\text{ML} = 0.67$

$p^* = \dfrac{\text{ML}}{\text{MP} + \text{ML}} = \dfrac{0.67}{1.50} = 0.4467$ which corresponds to $.13\sigma$, so he should order $\mu + .13\sigma = 375 + .13(20) = 377.6$, or 378 hot dogs.

b) Selling leftover hot dogs for \$0.50 leaves MP unchanged, but now $\text{ML} = 0.67 - 0.50 = 0.17$, so

$p^* = \dfrac{\text{ML}}{\text{MP} + \text{ML}} = \dfrac{0.17}{1.00} = .0.17$ which corresponds to $.95\sigma$, so he should order $\mu + .95\sigma = 375 + .95(20) = 394$ hot dogs.

17-14 $\text{MP} = 80$ $\quad\text{ML} = 135$

$p^* = \dfrac{\text{ML}}{\text{MP} + \text{ML}} = \dfrac{135}{215} = .6279,$ which corresponds to $-.33\sigma$, so they should prepare $\mu - .33\sigma = 190 - .33(32) = 179.44$ orders, or 90 chickens.

17-16 a) In a 50% tax bracket, to earn 9.43% after taxes, Bill must earn 18.86% before taxes.
18.86% of \$1,600,000 = \$301,760

b) Utility

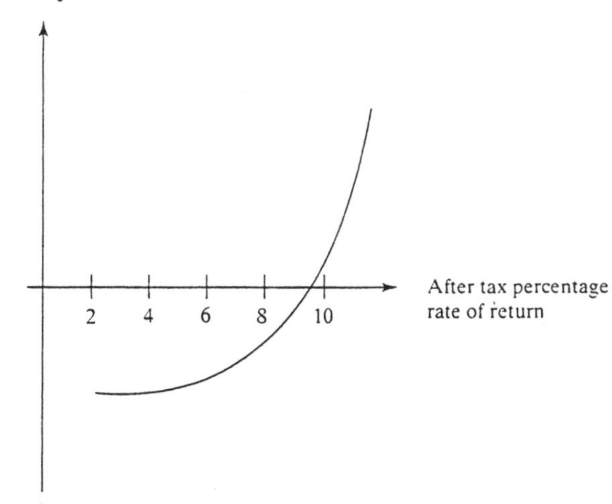

17-18

Change	−1000	−500	0	500	1000	1500	Expected utility
Utility	0.0	0.1	0.7	0.8	0.9	1.0	
P(change | 2-month option)	.05	.15	.15	.25	.35	.05	0.685
P(change | 4-month option)	0	.05	.05	.20	.30	.40	0.870
P(change | no purchase)	0	0	1	0	0	0	0.700

She should purchase the 4-month option.

17-20 $7/18 = .3889$, corresponding to 1.22σ, so $\sigma = 1/1.22 = .82$ month.
The probability that the upswing will last longer than six months is
$P(z > (6 - 8)/.82) = P(z > -2.44) = .9927$, which exceeds 95%. He should hire.

17-22 $4/14 = .2857$, corresponding to $.79\sigma$, so $\sigma = 12000/.79 = 15190$ miles.
Tax savings per mile $= .31 \times .34 = .1054$ so her breakeven mileage is
$12,250/.1054 = 116,224$ miles, corresponding to $z = \dfrac{116,224 - 120,000}{15,190} = -0.25$.
$P(z \geq -0.25) = .5987$ is the probability that the car will last long enough for Natalie to break even.

17-24 Broker A: $2/6 = .3333$, corresponding to $.97\sigma$, so $\sigma = 5/.97 = 5.155$.

Thus, $P(\text{price} \geq 60) = P\left(z \geq \dfrac{60 - 68}{5.155}\right) = P(z \geq -1.55) = .9394$

Broker B: $5/12 = .4167$, which corresponds to 1.38σ, so $\sigma = 5/1.38 = 3.623$.

Thus, $P(\text{price} \geq 60) = P\left(z \geq \dfrac{60 - 65}{3.623}\right) = Pz \geq -1.38) = .9162$

Since both of these are above .80, buy the stock.

17-26

a) Now the payoffs on the "operate by self, with snow-maker" branches become 95, 43, and −9, with an expected value of 43. Hence she should let the hotel operate the resort.

b) In this case, those three payoffs become 96, 48, and 0, with an expected value of 48. She should operate the resort by herself, using the snow-making equipment.

c) If we let the payoffs on those branches be $98 - 10x$, $58 - 50x$, and $18 - 90x$, then the expected value is $58 - 50x$. Setting this equal to the profit gained from letting the hotel chain operate the resort, we get

$$58 - 50x = 45$$
$$50x = 13$$
$$x = 13/50 = .26$$

Hence if the operating cost increases by 26%, she will be indifferent to those two strategies, since either will earn a profit of $45,000 for her.

17-28

```
LATE                                   LATE                              LATE
TIME  UTILITY               DECISION   TIME   UTILITY             PROB   TIME  UTILITY
====  =======               ========   ====   =======             ====   ====  =======
                                                          +---     20%    10      95
                                                          |
                    +--- BUS           15.50   82.50 ---( )--      50%    15      85
                    |                                     |
                    |                                     +---     30%    20      70
                    |
                    |                                     +---     80%    15      85
                    +--- WALK          16.00   82.00 ---( )
                    |                                     +---     20%    20      70
                    |
14.00   86.00 ---[ ]                                      +---     50%    10      95
                    |                                     |
                    +--- BIKE          14.00   86.00 ---( )--      40%    15      85
                    |                                     |
                    |                                     +---     10%    30      45
                    |
                    |                                     +---     30%    10      95
                    |                                     |
                    |                                     +---     45%    15      85
                    +--- CAR           15.25   83.25 ---( )
                                                          +---     15%    20      70
                                                          |
                                                          +---     10%    25      60
```

a) To minimize expected late time, Sam should ride his bike.

b) To maximize expected utility (test score), Sam should also ride his bike.

17-30 a) Let x be the number of print heads after which production is cheaper than purchase. Then,

$$\begin{aligned} 24x + 28000 &\leq 35x \\ 28000 &\leq 11x \\ 2545.5 &\leq x \end{aligned}$$

Now 2545 print heads corresponds to 2545/1.15 = 2213 units.

$$\text{P(demand} \geq 2213) = \text{P}\left(z \geq \frac{2213 - 3000}{700}\right) = \text{P}(z \geq -1.12) = .8686$$

b) Breakeven plus 1.5 standard deviations is 2213 + 1.5(700) = 3263 > μ, so the probability of being this far above breakeven is less than 50%. They should buy the modules.

17-32 a) ML = 14.70 MP = 26.95 − 14.70 = 12.25

$$p^* = \frac{\text{ML}}{\text{MP} + \text{ML}} = \frac{14.7}{26.95} = .545,$$ so he should order 44 tails (22 entrées).

b) Ordering 44 tails, his expected profit is

$$26.95[.07(18) + .09(19) + .11(20) + .16(21) + .57(22)] - 22(14.70) = \$244.44$$

Hence EVPI = 12.25[.07(18) + .09(19) + .11(20) + .16(21) + .20(22) + .15(23) + .14(24) + .08(25)] − 244.44 = \$21.88,

if requiring orders in advance doesn't change the demand distribution.

17-34 ML = 26.00 MP = 42.75 − 26.00 = 16.75

$$p^* = \frac{\text{ML}}{\text{MP} + \text{ML}} = \frac{26.00}{42.75} = .608,$$ so he should order 35 bags.

17-36 a) The three numbers at some nodes are the costs in parts b, d.i, and d.ii.

```
                CASE #1                  TRIAL                           CASE #2                  TRIAL
COST            DECISION    COST         RESULT   PROB   COST            DECISION   COST          RESULT   PROB    COST
====            ========    ====         ======   ====   ====            ========   ====          ======   ====    ====
                                                                                                                   95.0
                                                                   +-- SETTLE -------------------------------      120.0
                                                         95.0      I                                              165.0
                                                         120.0 --[ ]
           +-- SETTLE ---------------------------        136.5     I   GO TO                  +-- WIN ---   40%   82.5
           I                                                       +-- TRIAL -- 136.5 --( )
           I                                                                                  +-- LOSE --   60%   172.5
           I
           I                                                                                                       27.5
87.5       I                                                       +-- SETTLE -------------------------------      52.5
99.0 --[ ]                                               27.5      I                                               97.5
115.2      I                         +-- WIN ---  60%    30.0 --[ ]
           I                         I                   30.0      I   GO TO                  +-- WIN ---   80%    12.0
           I                         I                             +-- TRIAL --   30.0 --( )
           I   GO TO       87.5      I                                                        +-- LOSE --   20%   102.0
           +-- TRIAL --    99.0 --[ ]
                           115.2     I                                                                            177.5
                                     I                             +-- SETTLE -------------------------------     202.5
                                     I                   177.5     I                                              247.5
                                     +-- LOSE --  40%    202.5 --[ ]
                                                         243.0     I   GO TO                  +-- WIN ---   10%   162.0
                                                                   +-- TRIAL -- 243.0 --( )
                                                                                              +-- LOSE --   90%   252.0
```

b) He should take #1 to trial. If he wins #1, he should take #2 to trial; but, if he loses #1, he should settle #2 out of court.

c)

```
                MOCK TRIAL
COST            RESULT          PROB   ACTION          COST
====            ==========      ====   ======          ====
           +-- WIN #1   ----    60%    TRY BOTH        30.0
66.0 --( )
           +-- LOSE #1  ----    40%    SETTLE BOTH    120.0
```

He would pay up to 99 − 66 = 33, i.e., $33,000 for an absolutely reliable mock trial.

d) i. He would take #1 to trial, but settle #2 regardless of the outcome of the trial.
   ii. He would take #1 to trial, and take #2 to trial regardless of the outcome of the first trial.

17-38 ML = 21.50 − 19.95 = 1.55   MP = 43.95 − 21.50 = 22.45

$p^* = \frac{\text{ML}}{\text{MP} + \text{ML}} = \frac{1.55}{24.00} = .065$, so she should stock 25 suits.

17-40

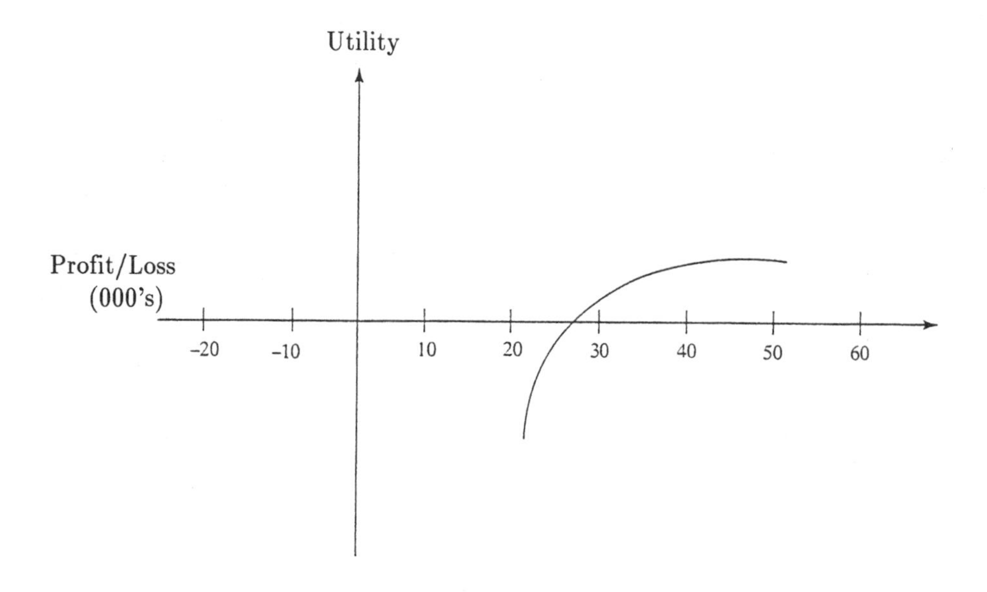

17-42 a) Assuming they take the analysis, we use Bayes' Theorem to find the revised probabilities:

Agency Rating	True Category	P(category)	P(rating | category)	P(rating & category)	P(category | rating)
	Poor	.25	.1	.025	.025/.25 = .10
A	Average	.45	.1	.045	.045/.25 = .18
	Good	.30	.6	.180	.180/.25 = .72
				P(A) = .250	
	Poor	.25	.2	.050	.05/.5 = .10
B	Average	.45	.8	.360	.36/.5 = .72
	Good	.30	.3	.090	.09/.5 = .18
				P(B) = .500	
	Poor	.25	.7	.175	.075/.25 = .70
C	Average	.45	.1	.045	.015/.25 = .18
	Good	.30	.1	.030	.030/.25 = .12
				P(C) = .250	

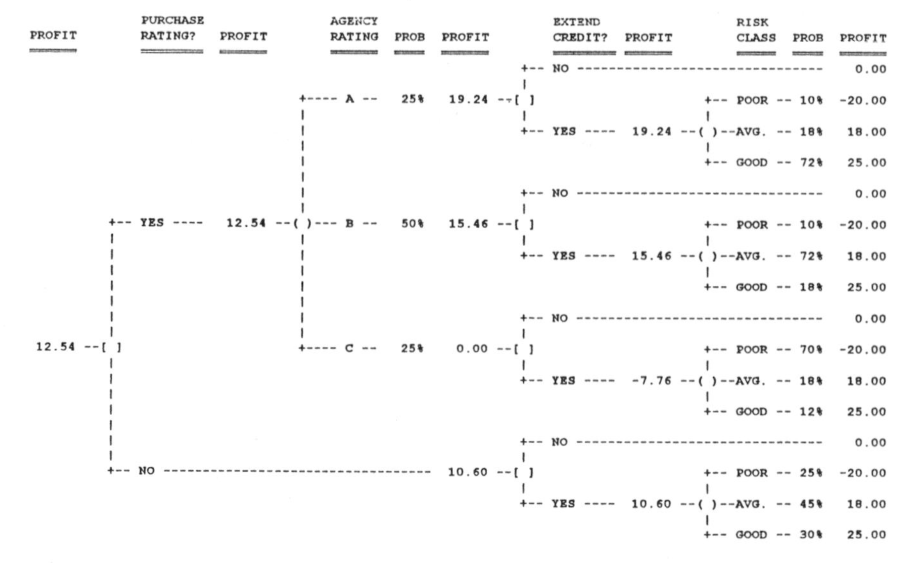

Since 12.54 − 10.60 = 1.94 > 0.75, they should purchase the credit rating.

b) Credit should be granted with an A or B rating, but not with a C rating.

c) 12.54 − 10.60 = 1.94, so they will pay at most $1940 for the rating.

d)

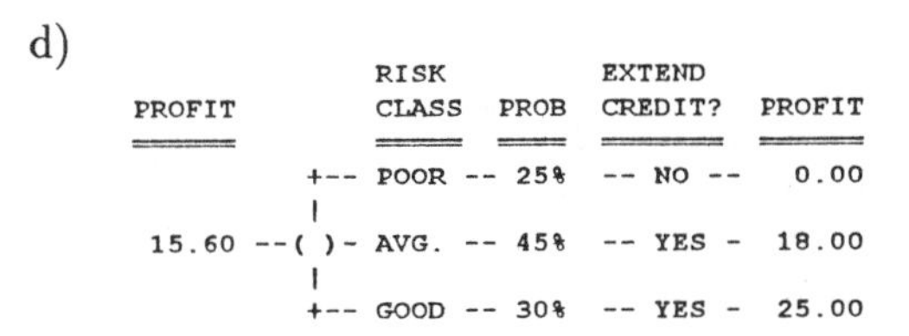

```
                     RISK               EXTEND
PROFIT               CLASS    PROB     CREDIT?    PROFIT
======               =====    ====     =======    ======
              +--    POOR --  25%      -- NO --     0.00
              |
15.60  --( )-        AVG. --  45%      -- YES -    18.00
              |
              +--    GOOD --  30%      -- YES -    25.00
```

15.60 − 10.60 = 5, so they would pay up to $5000 for an absolutely reliable credit report.

17-44 a) Expected profit = $80,000; therefore 1/2 of 80,000 = $40,000
$40,000 ÷ 500,000 = 8% return.
Enduro would accept, but the 8% return is below Steelfab's 9% cutoff.

b)

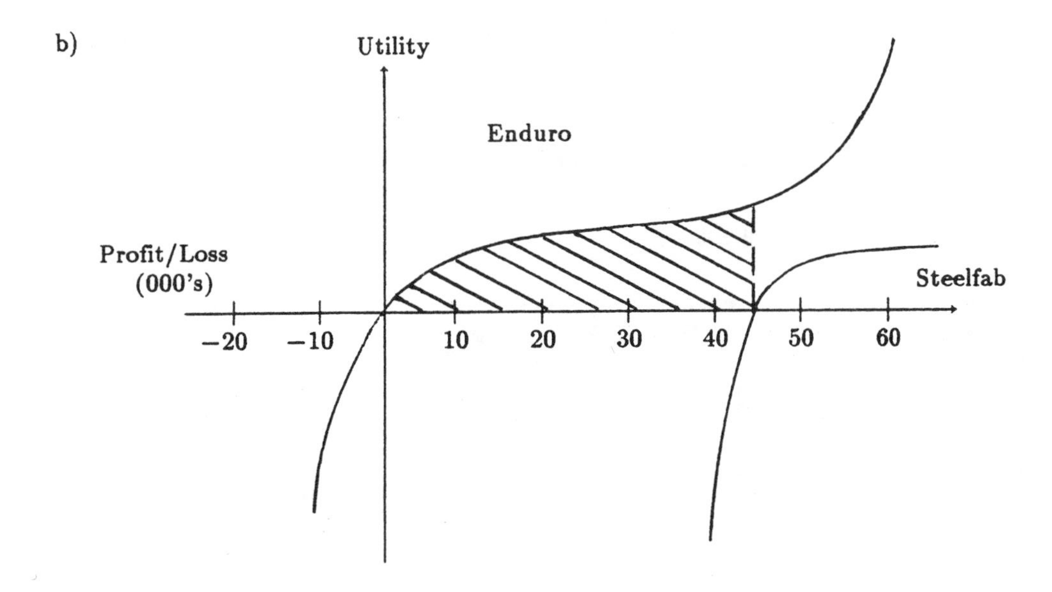

c) 55/500 = 11% return, which is acceptable to both. Steelfab would bid up to $611,111, where the $55,000 equals a 9% return.

17-46 a)

Number of games	Expected payoff
20	12600(.55) + 11000(.30) + 10600(.15) = $11,820
25	18000(.55) + 16200(.30) + 8500(.15) = $16,035
35	23000(.55) + 15000(.30) + 7100(.15) = $18,215

He should have 35 video games in the Cincinnati amusement center.

b) EVPI = 23000(.55) + 16200(.30) + 10600(.15) − 18215 = $885.

17-48 MP = selling price − cost = 1.50 − 0.70 = 0.80

ML = cost = 0.70

$$p^* = \frac{\text{ML}}{\text{MP} + \text{ML}} = \frac{0.70}{0.80 + 0.70} = 0.4667$$

Number of units	Probability of selling at least this many
500	.10 + .12 + .15 + .33 + .30 = 1.00
600	.12 + .15 + .33 + .30 = 0.90
700	.15 + .33 + .30 = 0.78
800	.33 + .30 = 0.63
900	.30 = 0.30

Since .63 > p^* > .30, Records and Tapes Unlimited should order 800 copies.

17-50 a)

Number of beds	Expected profit
50	41000(.2) + 52000(.3) + 65000(.5) = $56,300
75	− 12000(.2) + 68000(.3) + 80000(.5) = $58,000
50	− 53000(.2) − 24000(.3) + 117000(.5) = $40,700

He should build a 75-bed facility.

b) Expected profit with perfect information = 41000(.2) + 68000(.3) + 11700(.5) = $87,100

c) EVPI = 87100 − 58000 = $29,100

17-52 Since no salvage value is given, assume any unsold machines are worthless.

MP = selling price − cost = 89.95 − 75.50 = 14.45

ML = cost = 75.50

$$p^* = \frac{\text{ML}}{\text{MP} + \text{ML}} = \frac{75.50}{14.45 + 75.50} = 0.8394$$

Number of units	Probability of selling at least this many
15	.12 + .17 + .26 + .23 + .15 + .05 + .02 = 1.00
16	.17 + .26 + .23 + .15 + .05 + .02 = 0.88
17	.26 + .23 + .15 + .05 + .02 = 0.71

Since .88 > p^* > .71, Phones and More should order 16 phones.

17-54 a) 3150(1.08) = $3402

b.i) 100[0.25(25) + 0.50(35) + 0.25(50)] = $3625

ii) 525[0.25(0) + 0.50(35 − 30) + 0.25(50 − 30)] = $3937.50

c) They should sell now and use the proceeds to buy LEAPs, since that has the highest expected value.